Workshop on Biomedical Natural Language Processing 2010

(BioNLP 2010)

Held at ACL 2010

Uppsala, Sweden
15 July 2010

ISBN: 978-1-61738-810-1

Printed from e-media with permission by:

Curran Associates, Inc.
57 Morehouse Lane
Red Hook, NY 12571

Some format issues inherent in the e-media version may also appear in this print version.

Printed by Curran Associates, Inc. (2010)

For permission requests, please contact the Association for Computational Linguistics at the address below.

Association for Computational Linguistics
209 N. Eighth Street
Stroudsburg, Pennsylvania 18360

Phone: 1-570-476-8006
Fax: 1-570-476-0860

acl@aclweb.org

Additional copies of this publication are available from:

Curran Associates, Inc.
57 Morehouse Lane
Red Hook, NY 12571 USA
Phone: 845-758-0400
Fax: 845-758-2634
Email: curran@proceedings.com
Web: www.proceedings.com

BioNLP 2010: Year in review

Dina Demner-Fushman, K. Bretonnel Cohen,
Sophia Ananiadou, John Pestian, Jun'ichi Tsujii, and Bonnie Webber

Interest continues to increase in Biomedical Natural Language Processing, evidenced by the number of venues dedicated to BioNLP, the publication of a special issue of the *Journal of Biomedical Informatics* on Biomedical Natural Language Processing (Chapman and Cohen 2009), and the new and ongoing initiatives on BioNLP standards, centralized repositories, and community-wide evaluations. The latter include the third BioCreAtIvE evaluation (since 2003) to determine the state of the art in biomedical text mining and information extraction; the fourth i2b2 challenge on identifying concepts and relations in clinical notes; the CALBC project that plans to annotate several hundred thousand MEDLINE abstracts on immunology; the BioNLP 2009 shared tasks[1] that attracted 42 teams (of which 24 submitted their final results); workshops at ISMB, LREC, NAACL, and ACL; sessions at the AMIA summits and symposia; and the fourth international symposium on Semantic Mining in Biomedicine (SMBM).

The developments in BioNLP parallel key developing areas in medical informatics, including computerized clinical decision support, telemedicine, biosurveillance, personalized medicine, comparative effectiveness studies, and global health, as well as the emergence of clinical informatics as a branch of informatics. Personalized medicine has also been identified as key in translational and bioinformatics research. Other key areas in bioinformatics were high-throughput studies, literature mining, genetic privacy, environmental genetics, small molecules, pathways, and stem cell biology.

As in years past, authors have chosen the BioNLP workshop as a venue for presenting work that is innovative, novel, and challenging from an NLP perspective. The workshop received 34 submissions, of which nine were accepted as full papers and an additional twelve were accepted as posters. With very few exceptions, the submissions were of exceptional quality and we sincerely regret having to reject some of the good-quality work.

The themes in this years papers and posters cover complex NLP problems ranging from the foundations, such as a new approach to dealing with arguments of nominalizations (Kilicoglu et al. 2010), to high-level tasks, such as an approach to predicting breast cancer stage using social networking analysis (Jha and Elhadad 2010). Those who were waiting for the word sense disambiguation efforts to bear fruit will be glad to see the results of a graph-based WSD applied to concept-based summarization (Plaza et al. 2010). The growing maturity of the field continues to show in careful comparisons of available tools, for example, comparing widely-used syntactic parsers (Miwa et al. 2010). We also see re-use and expansion of the available collections, for example, the BioNLP 2009 event extraction collection (Vlachos 2010). In addition to event extraction (Björne et al. 2010, Ohta et al. 2010) and advanced methods for entity extraction (Liu et al. 2010), the program presents a method and tool for extraction of information about the expression of genes and their anatomical locations (Gerner et al. 2010).

Completing the program is work on expanding the set of methods for identifying negation and speculation, methods for detection of adverse reactions to drugs, corpus-based derivation of ontology for consequences of gene mutations, annotation methods and other timely topics presented in the poster session.

[1] http://www-tsujii.is.s.u-tokyo.ac.jp/GENIA/SharedTask/

Keynote: Text Mining and Intelligence

W. John Wilbur, MD, PhD

John Wilbur obtained a PhD in pure mathematics from the University of California at Davis and an MD from Loma Linda University. He is a Senior Investigator in the Computational Biology Branch of the National Center for Biotechnology Information which is located in the US National Library of Medicine. He is a principal investigator leading a research group in the study and development of statistical text processing algorithms. While at NCBI he has developed a number of algorithms that are used in the PubMed search engine including those for finding related documents, performing fuzzy phrase matching, and spell checking users queries.

Abstract

Humans are much more accurate at text mining than machines. Presumably this is because humans are more intelligent than machines. We argue that the way to narrow this gap is by more effective machine learning. One obvious difference between humans and machines is the large amounts of training data machines require for successful learning. We will discuss some novel ways of obtaining training data for machine learning. We will also discuss why humans appear to be different in their requirements for training data and what this may imply for the future of machine learning.

Acknowledgments

We are profoundly grateful to the authors who chose BioNLP from the smorgasbord of the enticing venues available this year. The authors willingness to share their work through BioNLP consistently makes the workshop noteworthy and stimulating. We are equally indebted to the program committee members (listed elsewhere in this volume) who produced at least two thorough reviews per paper on a tight review schedule and with an admirable level of insight. Finally, we acknowledge the gracious sponsorship of the Computational Medicine Center and Division of Biomedical Informatics, Cincinnati Childrens Hospital Medical Center.

References

Björne, Jari; Filip Ginter; Sampo Pyysalo; and Tapio Salakoski (2010) Scaling up event extraction: Targeting the entire PubMed. *BioNLP 2010.*

Chapman, Wendy; and K. Bretonnel Cohen (2009) Current issues in biomedical text mining and natural language processing. *Journal of Biomedical Informatics* 42(5):757-759.

Gerner, Martin; Goran Nenadic; and Casey M. Bergman (2010) An exploration of mining gene expression mentions and their anatomical locations from text. *BioNLP 2010.*

Jha, Mukund; and Noemie Elhadad (2010) Cancer stage prediction based on patient online discourse. *BioNLP 2010.*

Kilicoglu, Halil; Marcelo Fiszman; Graciela Rosemblat; Sean Marimpietri; and Thomas Rindflesch (2010) Arguments of nominals in semantic interpretation of biomedical text. *BioNLP 2010.*

Liu, Jingchen; Minlie Huang; and Xiaoyan Zhu (2010) Recognizing biomedical named entities using skip-chain conditional random fields. *BioNLP 2010.*

Miwa, Makoto; Sampo Pyysalo; Tadayoshi Hara; and Jun'ichi Tsujii (2010) A comparative study of syntactic parsers for event extraction. *BioNLP 2010.*

Ohta, Tomoko; Sampo Pyysalo; Makoto Miwa; Jing-Dong Kim; and Jun'ichi Tsujii (2010) Event extraction for post-translational modifications. *BioNLP 2010.*

Plaza, Laura; Mark Stevenson; and Alberto Diaz Esteban (2010) Improving summarization of biomedical documents using word sense disambiguation. *BioNLP 2010.*

Vlachos, Andreas (2010) Two strong baselines for the BioNLP 2009 event extraction task. *BioNLP 2010.*

BioNLP Sponsor:

Computational Medicine Center

Cincinnati Children's

University of Cincinnati

Organizers:

Kevin Bretonnel Cohen, University of Colorado School of Medicine
Dina Demner-Fushman, US National Library of Medicine
Sophia Ananiadou, University of Manchester and National Centre for Text Mining, UK
John Pestian, Computational Medical Center, University of Cincinnati, Cincinnati Children's Hospital Medical Center
Jun'ichi TsujiiUniversity of Tokyo and University of Manchester and National Centre for Text Mining, UK
Bonnie Webber,University of Edinburgh, UK

Program Committee:

Alan Aronson
Olivier Bodenreider
Bob Carpenter
Wendy Chapman
Aaron Cohen
Nigel Collier
Noemie Elhadad
Marcelo Fiszman
Kristofer Franzen
Jin-Dong Kim
Marc Light
Zhiyong Lu
Aurelie Neveol
Serguei Pakhomov
Thomas Rindflesch
Daniel Rubin
Hagit Shatkay
Larry Smith
Yuka Tateisi
Yoshimasa Tsuruoka
Karin Verspoor
Peter White
W. John Wilbur
Limsoon Wong
Hong Yu
Pierre Zweigenbaum

Invited Speaker:

W. John Wilbur, National Center for Biotechnology Information, US National Library of Medicine

Table of Contents

Two Strong Baselines for the BioNLP 2009 Event Extraction Task
Andreas Vlachos 1

Recognizing Biomedical Named Entities Using Skip-Chain Conditional Random Fields
Jingchen Liu, Minlie Huang and Xiaoyan Zhu 10

Event Extraction for Post-Translational Modifications
Tomoko Ohta, Sampo Pyysalo, Makoto Miwa, Jin-Dong Kim and Jun'ichi Tsujii 19

Scaling up Biomedical Event Extraction to the Entire PubMed
Jari Björne, Filip Ginter, Sampo Pyysalo, Jun'ichi Tsujii and Tapio Salakoski 28

A Comparative Study of Syntactic Parsers for Event Extraction
Makoto Miwa, Sampo Pyysalo, Tadayoshi Hara and Jun'ichi Tsujii 37

Arguments of Nominals in Semantic Interpretation of Biomedical Text
Halil Kilicoglu, Marcelo Fiszman, Graciela Rosemblat, Sean Marimpietri and Thomas Rindflesch
46

Improving Summarization of Biomedical Documents Using Word Sense Disambiguation
Laura Plaza, Mark Stevenson and Alberto Díaz 55

Cancer Stage Prediction Based on Patient Online Discourse
Mukund Jha and Noemie Elhadad 64

An Exploration of Mining Gene Expression Mentions and Their Anatomical Locations from Biomedical Text
Martin Gerner, Goran Nenadic and Casey M. Bergman 72

Exploring Surface-Level Heuristics for Negation and Speculation Discovery in Clinical Texts
Emilia Apostolova and Noriko Tomuro 81

Disease Mention Recognition with Specific Features
Md. Faisal Mahbub Chowdhury and Alberto Lavelli 83

Extraction of Disease-Treatment Semantic Relations from Biomedical Sentences
Oana Frunza and Diana Inkpen 91

Identifying the Information Structure of Scientific Abstracts: An Investigation of Three Different Schemes
Yufan Guo, Anna Korhonen, Maria Liakata, Ilona Silins, Lin Sun and Ulla Stenius 99

Reconstruction of Semantic Relationships from Their Projections in Biomolecular Domain
Juho Heimonen, Jari Björne and Tapio Salakoski 108

Towards Internet-Age Pharmacovigilance: Extracting Adverse Drug Reactions from User Posts in Health-Related Social Networks
Robert Leaman, Laura Wojtulewicz, Ryan Sullivan, Annie Skariah, Jian Yang and Graciela Gonzalez 117

Semantic Role Labeling of Gene Regulation Events: Preliminary Results
Roser Morante 126

Ontology-Based Extraction and Summarization of Protein Mutation Impact Information
Nona Naderi and René Witte 128

Extracting Distinctive Features of Swine (H1N1) Flu through Data Mining Clinical Documents
Heekyong Park and Jinwook Choi 130

Towards Event Extraction from Full Texts on Infectious Diseases
Sampo Pyysalo, Tomoko Ohta, Han-Cheol Cho, Dan Sullivan, Chunhong Mao, Bruno Sobral, Jun'ichi Tsujii and Sophia Ananiadou 132

Applying the TARSQI Toolkit to Augment Text Mining of EHRs
Amber Stubbs and Benjamin Harshfield 141

Integration of Static Relations to Enhance Event Extraction from Text
Sofie Van Landeghem, Sampo Pyysalo, Tomoko Ohta and Yves Van de Peer 144

Conference Program

Thursday, July 15, 2010

9:00–9:15 Opening Remarks

Session 1: Extraction

9:15–9:40 *Two Strong Baselines for the BioNLP 2009 Event Extraction Task*
Andreas Vlachos

9:40–10:05 *Recognizing Biomedical Named Entities Using Skip-Chain Conditional Random Fields*
Jingchen Liu, Minlie Huang and Xiaoyan Zhu

10:05–10:30 *Event Extraction for Post-Translational Modifications*
Tomoko Ohta, Sampo Pyysalo, Makoto Miwa, Jin-Dong Kim and Jun'ichi Tsujii

10:30–11:00 Morning coffee break

Session 2

11:–12:00 Keynote speaker, W. John Wilbur: Text Mining and Intelligence

12:05–12:30 *Scaling up Biomedical Event Extraction to the Entire PubMed*
Jari Björne, Filip Ginter, Sampo Pyysalo, Jun'ichi Tsujii and Tapio Salakoski

12:30–14:00 Lunch break

Thursday, July 15, 2010 (continued)

Session 3: Foundations

14:00–14:25 *A Comparative Study of Syntactic Parsers for Event Extraction*
Makoto Miwa, Sampo Pyysalo, Tadayoshi Hara and Jun'ichi Tsujii

14:25–14:50 *Arguments of Nominals in Semantic Interpretation of Biomedical Text*
Halil Kilicoglu, Marcelo Fiszman, Graciela Rosemblat, Sean Marimpietri and Thomas Rindflesch

Session 4: High-level tasks

14:50–15:15 *Improving Summarization of Biomedical Documents Using Word Sense Disambiguation*
Laura Plaza, Mark Stevenson and Alberto Díaz

15:30–16:00 Afternoon coffee break

Session 4: High-level tasks, continued

16:00–16:25 *Cancer Stage Prediction Based on Patient Online Discourse*
Mukund Jha and Noemie Elhadad

16:25–16:50 *An Exploration of Mining Gene Expression Mentions and Their Anatomical Locations from Biomedical Text*
Martin Gerner, Goran Nenadic and Casey M. Bergman

16:50–17:00 Poster Boaster session and Conclusions

17:00–17:30 Poster session

17:00–17:30 *Exploring Surface-Level Heuristics for Negation and Speculation Discovery in Clinical Texts*
Emilia Apostolova and Noriko Tomuro

17:00–17:30 *Disease Mention Recognition with Specific Features*
Md. Faisal Mahbub Chowdhury and Alberto Lavelli

17:00–17:30 *Extraction of Disease-Treatment Semantic Relations from Biomedical Sentences*
Oana Frunza and Diana Inkpen

Thursday, July 15, 2010 (continued)

17:00–17:30 *Identifying the Information Structure of Scientific Abstracts: An Investigation of Three Different Schemes*
Yufan Guo, Anna Korhonen, Maria Liakata, Ilona Silins, Lin Sun and Ulla Stenius

17:00–17:30 *Reconstruction of Semantic Relationships from Their Projections in Biomolecular Domain*
Juho Heimonen, Jari Björne and Tapio Salakoski

17:00–17:30 *Towards Internet-Age Pharmacovigilance: Extracting Adverse Drug Reactions from User Posts in Health-Related Social Networks*
Robert Leaman, Laura Wojtulewicz, Ryan Sullivan, Annie Skariah, Jian Yang and Graciela Gonzalez

17:00–17:30 *Semantic Role Labeling of Gene Regulation Events: Preliminary Results*
Roser Morante

17:00–17:30 *Ontology-Based Extraction and Summarization of Protein Mutation Impact Information*
Nona Naderi and René Witte

17:00–17:30 *Extracting Distinctive Features of Swine (H1N1) Flu through Data Mining Clinical Documents*
Heekyong Park and Jinwook Choi

17:00–17:30 *Towards Event Extraction from Full Texts on Infectious Diseases*
Sampo Pyysalo, Tomoko Ohta, Han-Cheol Cho, Dan Sullivan, Chunhong Mao, Bruno Sobral, Jun'ichi Tsujii and Sophia Ananiadou

17:00–17:30 *Applying the TARSQI Toolkit to Augment Text Mining of EHRs*
Amber Stubbs and Benjamin Harshfield

17:00–17:30 *Integration of Static Relations to Enhance Event Extraction from Text*
Sofie Van Landeghem, Sampo Pyysalo, Tomoko Ohta and Yves Van de Peer

Two strong baselines for the BioNLP 2009 event extraction task

Andreas Vlachos
Computer Laboratory
University of Cambridge
av308@cl.cam.ac.uk

Abstract

This paper presents two strong baselines for the BioNLP 2009 shared task on event extraction. First we re-implement a rule-based approach which allows us to explore the task and the effect of domain-adapted parsing on it. We then replace the rule-based component with support vector machine classifiers and achieve performance near the state-of-the-art without using any external resources. The good performances achieved and the relative simplicity of both approaches make them reproducible baselines. We conclude with suggestions for future work with respect to the task representation.

1 Introduction

The term *biomedical event extraction* is used to refer to tasks whose aim is the extraction of information beyond the entity level. It commonly involves recognizing actions and relations between one or more entities. The recent BioNLP 2009 shared task on event extraction (Kim et al., 2009) focused on a number of relations of varying complexity in which an event consisted of a trigger and one or more arguments. It attracted 24 submissions and provided a basis for system development. The performances ranged from 16% to 52% in F-score.

In this paper we describe two strong baseline approaches for the main task (described in Sec. 2) with a focus on annotation costs and reproducibility. Both approaches rely on a dictionary of lemmas associated with event types (Sec. 3). First we re-implement the rule-based approach of Vlachos et al. (2009) using resources provided in the shared task. While it is unlikely to reach the performance of approaches combining supervised machine learning, exploring its potential can highlight what annotated data is useful and its potential contribution to performance. Also, given its reliance on syntax, it allows us to assess the importance of syntactic parsing. Nevertheless, the performance achieved (35.39% F-score) is competitive with systems that used more annotated data and/or other resources (Sec. 5).

Building on the error analysis of the rule-based approach, we replace the rule-based component with support vector machine (SVM) classifiers trained on partial event annotation in the form of trigger-argument associations (Sec. 6). The use of a trainable classifier highlights issues concerning the suitability of the annotated data as training material. Using a simple feature representation and no external resources, the performance rises to 47.89% in F-score, which would have been second best in the shared task (Sec. 7). The error analysis suggests that future work on event extraction should look into different task representations which will allow more advanced models to demonstrate their potential (Sec. 8). Both systems shall become publically available.

2 Definition, datasets and resources

The BioNLP 2009 shared task focused on extraction of events involving proteins. Protein recognition was considered a given in order to focus the research efforts on the novel aspects of the task. Nine event types were defined in the main task, which can be broadly classified in two classes. Simple events, namely Gene_expression, Transcription, Protein_catabolism, Phosphorylation, Localization and Binding, which have proteins as their Theme argument and Regulation events, namely Positive_regulation, Negative_regulation and (unspecified) Regulation which have an obligatory Theme argument and an optional Cause argument which can be either a protein or another event. Every event has a trigger which is a contiguous textual string that can span over one or more tokens, as well as a part of a token. Triggers and arguments can be shared across events and

Proceedings of the 2010 Workshop on Biomedical Natural Language Processing, ACL 2010, pages 1–9,
Uppsala, Sweden, 15 July 2010.

ID	type	trigger	Theme	Cause
E1	Neg_reg	suppressed	E2	
E2	Pos_reg	induced	E3	gp41
E3	Gene_exp	production	IL-10	

Table 1: Shared task example annotation.

the same textual string can be a trigger for events of different types. In an example demonstrating the complexity of the task: "...SQ 22536 suppressed **gp41**-induced **IL-10** production in monocytes." Participating systems, given the two proteins (in bold), need to generate the three appropriately nested events of Table 1.

While event components can reside in different sentences, we focus on events that are contained in a single sentence. Participants were not provided with resources to develop anaphora resolution components and the anaphoric phenomena involved were rather complex, as we observed in Vlachos et al. (2009). Extraction of events involving anaphoric relations inside a single sentence is still possible but it is likely to require rather complex patterns to be extracted.

The shared task involved three datasets, training, development and test, which consisted of 800, 150 and 260 abstracts respectively taken from the GENIA event corpus. Their annotation was tailored to the shared task definition. A resource made available and used by the majority of the systems was the output of four syntactic parsers:

- Bikel's (2004) re-implementation of Collins' parsing model. This parser was trained on newswire data exclusively.
- The re-ranking parser of Charniak & Johnson adapted to the biomedical domain (McClosky and Charniak, 2008). The in-domain, part-of-speech (PoS) tagger was trained on the GENIA corpus (Kim et al., 2003) and the self-training of the re-ranking module used a part of the GENIA treebank as development data.
- The C&C Combinatory Categorial Grammar (CCG) parser adapted to the biomedical domain (Rimell and Clark, 2009). The PoS tagger was trained on the GENIA corpus, while 1,000 sentences were annotated with lexical categories and added to the training data of the CCG supertagger and 600 sentences of the BioInfer corpus (Pyysalo et al., 2007) were used for parameter tuning.
- The GDep dependency parser trained for the biomedical domain in the experiments of Miyao et al. (2008). This parser was trained for the biomedical domain using the GENIA treebank.

The native Penn TreeBank output of Bikel's and McClosky's parser was converted to the Stanford Dependency (SD) collapsed dependency format (de Marneffe and Manning, 2008). The output of the CCG parser was also converted to the same dependency format, while the output of GDep was provided in a different dependency format used for the dependency parsing CoNLL 2007 shared task. From the description above, it is clear that the various parsers have different levels of adaptation to the biomedical domain. While it is difficult to assess quantitatively the actual annotation effort involved, it is possible to make some comparisons. Bikel's parser was not adapted to the domain, therefore it would be the cheapest one to deploy. McClosky and CCG used in-domain corpora annotated with PoS tags for training, while the latter using some additional annotation for lexical categories. Furthermore, they were tuned using in-domain syntactic treebanks. Therefore, they represent a more expensive option in terms of annotation cost. Finally, GDep was trained using an in-domain treebanked corpus, thus representing the alternative with the highest annotation cost.

3 Trigger extraction

We perform trigger identification using a dictionary of lemmas associated with the event type they indicate. The underlying assumption is that a particular lemma has the same semantic content in every occurrence, which results in extracting all of its occurrences as triggers of the same event type. This is clearly an over-simplification, but the restricted domain and the task definition alleviates most of the problems caused. For each lemma in the dictionary, we extract all its occurrences in the text as triggers, therefore over-generating, since not all occurrences denote a biomedical event. This can be either because they are not connected with appropriate arguments or because they are used with a sense irrelevant to the task. Both issues are being resolved at the argument identification stage since superfluous triggers should not receive arguments and not form events.

The one-sense-per-term assumption is further challenged by the fact that occurrences of the same term can denote events of different types. For example, "expression" is used as a trigger of four different event types in the training data, namely Gene_expression, Transcription, Localization and Positive_regulation. While it can be argued that in some cases this is due to annotation inconsistencies, it is generally accepted that context can alter the semantics of a token. In order to ameliorate this problem, we define the concept of *light triggers* in analogy analogy to *light verbs*. The latter are verbs whose semantics are lost when occurring in particular constructions, e.g. "make" as in "make mistakes". In the shared task, some lemmas commonly associated with a particular event type, when modified by a term associated with a different event type, denote events of the type of their modifier instead of their own. For example, "regulation" generally denotes Regulation events, unless it has a modifier of a different event type, e.g. "positive". In these cases, "regulation" becomes part of a multi-token Positive_regulation trigger (e.g. "positive regulation"). However, if the actual tokens are not adjacent, only "regulation" is annotated as a Positive_regulation trigger, which is due to the requirement that triggers are contiguous textual strings. We refer to lemmas exhibiting this behaviour as *light triggers*. Additionally, we observe that some lemmas triggered events only when modified by another lemma associated with an event type. For example, "activity" when occurring without a modifier is not considered a trigger of any event, however, when modified by "binding" then it becomes a Binding event trigger. We refer to lemmas exhibiting this behaviour as *ultra-light triggers*.[1]

In order to construct the dictionary of terms with their associated event types we use the trigger annotation from the training data, but we argue that such information could be obtained from domain experts. First, we remove the triggers encountered only once in the data in order to avoid processing non-indicative triggers. Then, we lemmatize them with *morpha* (Minnen et al., 2001). We remove prepositions and other stopwords from multi-token triggers such as "in response to" and "have a prominent increase" in order to keep only the terms denoting the event type. Then, using the single-token triggers only, we associate each lemma with its most common event type. In cases where a lemma consistently generates more that one event trigger of different types (typically one of the Simple event class and one of the Regulation class, we associate the lemma with all the relevant event types. For example, "overexpress" consistently denotes Gene_expression and Positive_regulation events. The last token of each multi-token trigger becomes a *light* trigger. Finally, if a lemma is encountered as part of a multi-token trigger of a different event type more often than with the event type associated with it as a single-token trigger, then it becomes an *ultra-light* trigger. We avoid stemming because suffixes distinguish lemmas in an important way with respect to the task. For example, "activation" denotes Positive_regulation events, while "activity" is an *ultra-light* trigger. We only keep lemmas associated at least four times with a particular event type, since below that threshold the annotation was rather inconsistent.

During testing, we attempt to match each token with one of the lemmas associated with an event type. We perform this by relaxing the matching successively, using the token lemma first and if no match is found allowing a partial match in order to deal with particles (e.g. so that "co-express" matches "express"). This process returns single-token triggers, some of which are processed further in case they are light or ultra-light using syntactic dependencies in the following stage.

4 Rule-based argument identification

In this stage, we connect the triggers extracted with appropriate arguments using rules defined with the Stanford dependency (SD) scheme (de Marneffe and Manning, 2008). We re-implement the set of rules of Vlachos et al. (2009) using the syntactic parsing resources provided by the organizers for the development data. Rule-based systems need annotated data for tuning, but unlike their supervised machine learning-based counterparts they do not learn parameters from it, thus requiring less annotated data. We consider this to be the main advantage of rule-based systems and to demonstrate this point we explicitly avoid using the training data provided. The rules define syntactic dependency paths that connect tokens containing triggers (*trigger-tokens*) with tokens containing their arguments (*arg-tokens*). For multi-token protein names, it is sufficient that a path reaches any of its tokens. For Regulation event

[1] Kilicoglu and Bergler (2009) made similar observations on the lemma "activity" without formalizing them.

class triggers we consider as *arg-tokens* not only tokens containing (parts of) protein names but also the *trigger-tokens* found in the same sentence. The rules defined are the following:

- If a *trigger-token* is the governor of an *arg-token* in subject relation (*subj*), then the latter is identified as the Theme argument of the former, e.g. "**Stat1** expresses". The only exception to this rule is that when the trigger denotes Regulation class events and the nominal subject relation (*nsubj*) is observed, the *arg-token* is identified as a Cause argument, e.g. "**gp41** induces".
- If a *trigger-token* is the governor of an *arg-token* in a prepositional relation, then the latter is identified as the Theme argument of the former, e.g. "expression of **Stat1**".
- If a *trigger-token* is the governor of an *arg-token* in modifier relation then the latter is identified as the Theme argument of the former, e.g. "**Stat1** expression". We restrict the definition of the modifier relation to subsume only the following relations: adjectival modifier (*amod*), infinitival modifier (*infmod*), participial modifier (*partmod*), adverbial modifier (*advmod*), relative clause modifier (*rcmod*), quantifier modifier (*quantmod*), temporal modifier (*tmod*) and noun compound modifier (*nn*) relations. This restriction is placed in order to avoid matches irrelevant to the task.
- If a *trigger-token* is the governor of an *arg-token* in object relation (*obj*) then the latter is identified as the Theme argument, e.g. "SQ 22536 suppressed **gp41**".
- If a Regulation event class trigger and a protein name are found in the same token, then the protein name is identified as the Cause argument, e.g. "**gp41**-induced".

A pre-processing step taken was to propagate modifier and prepositional relations over tokens that were co-ordinated or in an appositive relation. This was necessary since the SD output provided by the organizers is in the collapsed format, which treats co-ordinated tokens asymmetrically without propagating their dependencies.[2]

For each Simple or Binding trigger-argument pair, we generate a single event with the argument marked as Theme. This approach is expected to deal adequately with all event types except for Binding, which can have multiple themes. We generate Regulation events for trigger-argument pairs whose argument is a protein name or a trigger that has an already formed event. Since Regulation events can have other Regulation events as Themes or Causes, we repeat this process until no more events can be formed. Finally, at this stage we generate the required Regulation class event for triggers that consistently denote two events.

5 Rule-based system results

We report our results using the approximate span matching/approximate recursive matching variant of the evaluation. This variant allows for an event to be considered extracted correctly if its trigger is extracted with span within an one-token extension of the correct trigger span. Also in the case of nested events, events below the top-level need only their Theme argument to be correctly identified so that the top-level event is considered correct. This evaluation variant was used as the primary performance criterion in the shared task.

We first compared the performances obtained using the output of the different parsers provided by the organizers on the development data. The best F-score was achieved using McClosky (39.66%), followed by CCG (38.73%) and Bikel (36.97%). As expected, the overall performance correlates roughly with the adaptation cost involved in the development of these parsers as described in Section 2. Bikel, which is essentially unadapted, has the worst performance overall, but it would have been the cheapest to deploy. While this can be viewed as a task-based parser comparison, similar to the experiments of Miyao et al. (2008), one should be careful with the interpretation of the results. As pointed out by the authors, this type of evaluation cannot substitute a parsing evaluation against an appropriately annotated corpus since in the context of a given task only some aspects of parsing are likely to be relevant. Furthermore, in our experiments we are are not using the native output of the parsers but its conversion to the SD format. Therefore unavoidably we evaluate the conversion as well as the parsing. For this reason we avoided using the output of GDep which was not provided in this format.

Examining the lists of false positives and false negatives on the system using the McClosky

[2]The organizers re-generated the dependencies in the propagation format but we avoid using them in order to be able to compare against the shared task participants.

parser, we observe that the most common triggers of events not extracted correctly had lemmas that were included in the dictionary, such as "binding", "expression", "induction" and "activation". This suggests that most event extraction errors are due to argument identification and that using a dictionary for trigger extraction is sufficient, despite the rather strong assumptions it is based upon. Disabling the processing of light and ultra-light triggers, the performance on the development data drops to 39.28%, the main reason being the decreased recall in Binding events.

Based on the comparison performed on the development data, we run our system using the McClosky parser on the test data (Table 2). The overall performance achieved (35.39%) is relatively close to the one obtained on the development set (4% lower). This is important since rule-based systems are prone to overfitting their development data due to the way they are built. Compared to the performances achieved by the shared task participants, the system presented would be ranked seventh in overall performance. We believe this is a strong result, since it surpasses systems that used supervised machine learning methods taking advantage of the development and the training data. Restricting the comparison to rule-based systems, it would have the second best performance out of nine such systems, most of which used external knowledge sources in order to improve their performance. The best rule-based system (Kilicoglu and Bergler, 2009) had overall performance of 44.62% in F-score, ranking third overall. The main difference is that it used a much larger set of lexicalized rules (27) which were extracted using the training data. Also, heuristics were employed in order to correct syntactic parsing errors (Schuman and Bergler, 2006). While the benefits from these additional processing steps are indisputable, they involved a lot of manual work, both for rule construction as well as for the annotation of the data used to extract the rules. We argue that these performance benefits could be obtained using machine learning methods aimed at ameliorating the argument identification stage. Compared to the rule-based approach of Vlachos et al. (2009), the performance is improved substantially. The main difference between that system and the one presented here is that the former uses the domain-independent RASP parser (Briscoe et al., 2006). While its performance was reasonable (it was ranked 10th overall, 30.80% F-score), these results lag behind those reported here. Note that a direct comparison using the output of RASP is not possible since the latter uses its own syntactic dependency scheme and there is no lossless conversion to the SD scheme.

Overall, the results of this section demonstrate that the use of domain-adapted parsing is beneficial to event extraction. This is not surprising since the system presented depends heavily on the parsing output. We argue that the annotation cost of this adaptation is a good investment because, unlike the task-specific training data, improved syntactic parsing is likely to be useful for other event extraction tasks, or even other IE tasks, e.g. anaphora resolution. Therefore, we suggest that domain-adaptation of syntactic parsing should be considered first, especially in tasks that are heavily dependent on it.

6 Improving argument identification with partial annotation and support vector machines

In this section, we present an approach to argument identification which attempts to overcome the drawbacks of the rule-based approach. Following the trigger extraction stage, for each trigger combined with each of its candidate arguments we create a classification instance. The classification task is to assign the correct argument type to the instance. Therefore, we construct a binary classifier which determines whether a protein name is the Theme argument of a Simple or a Binding trigger (*ThemePositive* or *ThemeNegative*) and a ternary classifier which determines whether a protein name or another trigger (and as consequence its associated events) is the Theme or the Cause argument of a Regulation trigger (*RegThemePositive*, *RegCausePositive*, *RegNegative*).

In order to acquire labeled instances for training, we decompose the gold standard (GS) events into multiple events with single arguments. In cases of events being arguments to Regulation events, the former are replaced by their triggers. We match the triggers extracted with those included in the gold standard, ignoring the event type annotation. Since we identify single-token triggers, we replicate the approximate span matching used in evaluation in order to achieve better coverage. If the instance being considered has a Simple or a Binding trigger, and if the pair is in-

	Rules (MC)			SVM (MC+CCG)		
Event Type/Class	recall	precision	F-score	recall	precision	F-score
Gene_expression	46.54	78.50	58.43	61.63	82.26	70.47
Transcription	26.28	28.57	27.38	29.93	62.12	40.39
Protein_catabolism	28.57	100.00	44.44	42.86	85.71	57.14
Phosphorylation	65.19	82.24	72.73	78.52	91.38	84.46
Localization	32.18	88.89	47.26	40.80	95.95	57.26
Simple (total)	43.99	71.43	54.45	56.60	83.21	67.37
Binding	20.46	38.17	26.64	29.11	45.29	35.44
Regulation	15.81	23.47	18.89	23.71	39.20	29.55
Positive	21.16	33.02	25.79	37.03	43.65	40.07
Negative	17.15	29.41	21.67	30.34	40.35	34.64
Regulation (total)	19.30	30.47	23.63	33.15	42.32	37.18
Total	28.60	46.40	35.39	41.42	56.76	47.89

Table 2: Performance of the rule-based and the SVM-based systems on the test data. Each horizontal corresponds to an event type or class. Binding events are not included in the Simple class aggregate performance because they can have multiple Themes.

cluded in the GS then it is labeled as *ThemePositive*, else it labeled as *ThemeNegative*. If the instance being considered has a Regulation trigger that has been matched with a GS trigger, and if its argument is a protein name and their pair is included in the GS then it is labeled according to the latter (*Reg*{*Theme/Cause*}*Positive*), else, if not found in the GS it is labeled as *RegThemeNegative*. The same process is followed if the argument is an event trigger which has been matched with a GS trigger. We consider only Regulation triggers that are matched in the GS in order to avoid valid *RegCausePositive* instances being labeled as *RegNegative*. Recall that the Cause argument is optional, while the Theme is obligatory for Regulation events. This means that if an appropriate Theme argument is not present, then it is possible that a Cause argument that is present is not annotated. Similarly, when considering event triggers as arguments, we acquire labels only for instances involving triggers that were annotated in the GS. Since triggers without an appropriate Theme are not annotated in the gold standard, it is possible that a valid *RegThemePositive* or *RegCausePositive* is labeled as *RegNegative* instance not because of the actual relation between the trigger and the argument but because the argument did not have an appropriate Theme present. In the example mentioned in Sec. 2, if "IL-10" was replaced by "protein" then none of the events would be annotated. We argue that a human annotator would produce these annotations implicitly, and that this partial (with respect to the task definition) annotation scheme allows the encoding of this information in a more flexible way. Also, this is likely to be a more efficient way to use the annotation time, since annotators would be requested to annotate pre-determined trigger-argument pairs instead of searching for events from scratch, given only the protein name annotation.

For training data generation we consider the triggers extracted using the dictionary instead of those in the GS. This process is certain to introduce some noise as some triggers might be omitted due to limited dictionary coverage. If the event type determined by the dictionary is incorrect, this is unlikely to affect the argument identification process, since the latter is dependent on the lemma of the trigger rather than its type. For example, the Theme argument of the trigger "expression" is unlikely to depend on whether the event denoted is Gene_expression or Transcription.

The labeled instance acquisition process described results in 9,699 binary and 10,541 ternary labels compared to 6,607 triggers and 9,597 events annotated in the training data provided. However, it must be pointed out that in the shared task annotation scheme negative instances are annotated implicitly, i.e. non-events are not annotated. If we consider only the positive instances, then the annotation scheme describeed results in 3,517 *ThemePositive* and 3,933 *Reg*{*Theme/Cause*}*Positive* instances, which are simpler since they do not need require textual span and event type specification.

For feature extraction, we find the shortest dependency path connecting each trigger-argument pair using Dijkstra's algorithm. We allow paths to follow either dependency direction by incorporating the direction in the dependency labels. Apart from the dependency path, we extract as features the trigger-token, the trigger event type and the argument type (event type if the argument is a trigger or *Entity* in case of protein names). We filtered the training set considering only instances in which the trigger was at a maximum distance of 4 dependencies away from the argument, since longer paths were too sparse to be useful in classifying unseen instances. At classification time, we consider as {*Theme/Reg*}*Negative* any instances in which the dependency path has not been encountered in the training data, as well as instances without a dependency path connecting trigger and argument. This is necessary in order to avoid instances being classified only on the basis of the trigger-token and the argument type. After the classifier has assigned labels to the trigger-argument pairs, we construct events as described in Sec. 4. In cases where it is unclear (to the classifier) which is the trigger and which is the argument in a given pair of Regulation event triggers the process can result in cyclic dependencies. We resolve them using the confidence of the classifier for each decision by removing the least confident *RegThemePositive* or *RegCausePositive* assignment.

7 SVM-based system results

In our experiments we used the LIBSVM toolkit (Chang and Lin, 2001) which provides an implementation of Support Vector Machines with various kernels and uses the one-against-one scheme for multiclass problems. In all experiments, the Gaussian kernel was used in order to capture potential non-linear feature combinations, e.g. cases where the combination of dependency path and trigger-token would result in a different decision rather than each of them independently. Preliminary experiments with the linear kernel confirmed this expectation.

We focused on using the output of the two domain-adapted parsers, namely CCG and McClosky. The reason for this is that, as argued in Sec. 5, given the importance of syntactic parsing to event extraction one should consider domain adaptation of syntactic parsing before developing task-specific training resources. We first compared the performances obtained using the output of the different parsers provided by the organizers using the development data. The main observation is that, using either parser, the results are much improved compared to those reported in Sec. 5, by approximately eight percentage points in F-score in either case (46.49% and 47.40% F-score for CCG and McClosky respectively). Most of the improvement is due to higher recall, suggesting that the argument identification component is able to learn patterns that are relevant to the task. Overall, using the output of CCG results in higher precision, while McClosky results in higher recall. These parsers have different theoretical foundations, therefore they are expected to make different errors. In an effort to take advantage of both parsers simultaneously, we combined them by adding for each trigger-argument pair the dependency paths extracted by both parsers. This improved performance further to 49.35% F-score.

We then run the system combining the two parsers on the test data, obtaining the results presented in Table 2. Overall, the system presented would have the second best performance in the shared task achieving 41.42%/56.76%/47.89% in Recall/Precision/F-score. The top system (Bjorne et al., 2009) achieved 46.73%/58.48%/51.95% (R/P/F). It followed a machine learning approach to trigger extraction which, while it is likely to be responsible for the performance difference observed when compared to the other participating systems, requires explicit trigger annotation, thus being more expensive. Furthermore, we argue that the data provided by the organizers are not suitable to train a trigger extractor, since only triggers participating in events are annotated, and semantically valid triggers without appropriate arguments present are ignored. We hypothesize that this is the reason the authors had to adjust the decisions of their SVM classifiers.

The second best system (Buyko et al., 2009) achieved 45.82%/47.52%/46.66% (R/P/F) using many external knowledge sources such as the Gene Ontology Annotation database, the Universal Protein Resourceand the Medical Subject Headings thesaurus. While the use of these resources and their successful usage is commendable, we believe it is important that the system presented achieves comparable performance using fewer resources.

Furthermore, joint inference models such as

Markov Logic Networks were applied to the BioNLP 2009 event extraction shared task by Riedel et al. (2009) and were ranked fourth. This result was improved upon recently by Poon and Vanderwende (2010) who achieved 50% F-score, 2.11 percentage points better than the result achieved in this work. Such models have great potential for event extraction and we believe that they can benefit from the insights presented here. Finally, despite the fact that we used the same experimental setup as the shared task participants, we do not consider our results are directly comparable to theirs since we did not work under the same time constraints and we profited from their experiences.

8 Discussion

Our error analysis on the output of the best system on the development data discouraged us from pursuing further improvements. Echoing the observations of Buyko et al. (2009), we found that annotation inconsistency was affecting our results significantly. In many cases the event triggers annotated in the development data were rather misleading, e.g. "negative" as a Gene_expression event trigger (abstract 8622883), "increase the stability" as a Positive_regulation event trigger (abstract 8626752), "disappearance" as a Binding event trigger (abstract 10455128). Finally, some events were ignored by the annotation, such as "regulation of **thymidine kinase**" (abstract 8622883).

An additional complication is that events that are annotated due to anaphoric linking can have a disproportionate effect on the scores. In an example from abstract 9794375: "CD3, **CD2**, and **CD28** are functionally distinct receptors on T lymphocytes. Engagement of any of these receptors induces the rapid tyrosine phosphorylation of a shared group of intracellular signaling proteins, including **Vav**, **Cbl**, p85 phosphoinositide 3-kinase, and the **Src** family kinases **Lck** and **Fyn**." Failing to recognize the anaphoric Binding events involving proteins "CD2" and "CD28", an otherwise perfect system would receive two false negatives for the Binding events, eight false negatives for the missing Positive_regulation events due to the missing Causes and four false positives for the incomplete Positive_regulation events extracted.

Despite this criticism, we believe that the BioNLP 2009 shared task on event extraction was a big step forward for biomedical information extraction and we are grateful to the organizers for the effort and resources provided, without which the research presented here would not have been possible. The performances achieved in the main Task1 ranged from 16% to 52% in F-score, suggesting improvements in task definition, data annotation and participating systems compared to previous community-wide efforts. Indicatively, in the protein-protein interaction pair subtask of BioCreative II (Krallinger et al., 2008) the annotated datasets provided were produced by extracting curation information from relevant databases. This meant that there was no text-bound annotation, thus making the application and evaluation of existing NLP techniques difficult, resulting in rather low performances. The best performance achieved was 29% in F-score, while many of the teams scored below 10%.

However, we believe that future work should look at improving the annotation in order to be able to assess the progress in the systems developed. In particular, we argue that we should move towards a dependency-based representation, similar to the one introduced by Surdeanu et al. (2008) for joint syntactic parsing and semantic role labeling. Such representation can express the nested nature of the events and evaluate the dependencies between them directly. Furthermore, given the importance of syntactic parsing via syntactic dependencies to event extraction, it would be interesting to see how performing these tasks jointly would help improve the performance. A dependency-based representation would also allow for non-contiguous event components, as well as more complex phenomena such as the *light triggers* discussed earlier.

9 Conclusions

In this paper we focused on the BioNLP 2009 shared task on event extraction. We developed two systems, a rule-based one that does not require training data and a SVM-based one which achieves near state-of-the-art performance. The good performances achieved and their reliance on shared task resources exclusively makes them reproducible and strong baselines for future work. Furthermore, we demonstrated the importance of domain adaptation of syntactic parsing for event extraction. Finally, based on our error analysis we suggest future directions for event extraction with respect to the task representation.

Acknowledgements

The author would like to thank Ted Briscoe, Mark Craven and the three anonymous reviewers for their feedback.

References

Daniel M. Bikel. 2004. Intricacies of Collins' parsing model. *Computational Linguistics*, 30(4):479–511.

Jari Bjorne, Juho Heimonen, Filip Ginter, Antti Airola, Tapio Pahikkala, and Tapio Salakoski. 2009. Extracting complex biological events with rich graph-based feature sets. In *Proceedings of the BioNLP'09 Shared Task on Event Extraction*, pages 10–18.

Ted Briscoe, John Carroll, and Rebecca Watson. 2006. The second release of the RASP system. In *Proceedings of the COLING/ACL Interactive presentation sessions*, pages 77–80, Morristown, NJ, USA. Association for Computational Linguistics.

Ekaterina Buyko, Erik Faessler, Joachim Wermter, and Udo Hahn. 2009. Event extraction from trimmed dependency graphs. In *Proceedings of the BioNLP 2009 Workshop Companion Volume for Shared Task*, pages 19–27.

Chih-Chung Chang and Chih-Jen Lin, 2001. *LIBSVM: a library for support vector machines*. Software available at http://www.csie.ntu.edu.tw/~cjlin/libsvm.

Marie-Catherine de Marneffe and Christopher D. Manning. 2008. The stanford typed dependencies representation. In *CrossParser '08: Coling 2008: Proceedings of the Workshop on Cross-Framework and Cross-Domain Parser Evaluation*, pages 1–8.

Halil Kilicoglu and Sabine Bergler. 2009. Syntactic dependency based heuristics for biological event extraction. In *Proceedings of the BioNLP 2009 Workshop Companion Volume for Shared Task*, pages 119–127.

Jin-Dong Kim, Tomoko Ohta, Yuka Tateisi, and Jun'ichi Tsujii. 2003. GENIA corpus - a semantically annotated corpus for bio-textmining. In *Bioinformatics*, volume 19, Suppl. 1, pages 180–182.

Jin-Dong Kim, Tomoko Ohta, Sampo Pyysalo, Yoshinobu Kano, and Jun'ichi Tsujii. 2009. Overview of BioNLP'09 shared task on event extraction. In *BioNLP '09: Proceedings of the Workshop on BioNLP*, pages 1–9.

Martin Krallinger, Florian Leitner, Carlos Rodriguez-Penagos, and Alfonso Valencia. 2008. Overview of the protein-protein interaction annotation extraction task of BioCreative II. *Genome Biology*, 9(Suppl 2):S4.

David McClosky and Eugene Charniak. 2008. Self-training for biomedical parsing. In *Proceedings of the 46th Annual Meeting of the Association of Computational Linguistics: Human Language Technologies*, pages 101–104.

Guido Minnen, John Carroll, and Darren Pearce. 2001. Applied morphological processing of English. *Natural Language Engineering*, 7(3):207–223.

Yusuke Miyao, Rune Sætre, Kenji Sagae, Takuya Matsuzaki, and Jun'ichi Tsujii. 2008. Task-oriented evaluation of syntactic parsers and their representations. In *Proceedings of the 46th Annual Meeting of the Association of Computational Linguistics: Human Language Technologies*, pages 46–54.

Hoifung Poon and Lucy Vanderwende. 2010. Joint inference for knowledge extraction from biomedical literature. In *Proceedings of the North American Chapter of the Association for Computational Linguistics - Human Language Technologies 2010 conference*.

Sampo Pyysalo, Filip Ginter, Juho Heimonen, Jari Bjorne, Jorma Boberg, Jouni Jarvinen, and Tapio Salakoski. 2007. BioInfer: a corpus for information extraction in the biomedical domain. *BMC Bioinformatics*, 8(1):50+.

Sebastian Riedel, Hong-Woo Chun, Toshihisa Takagi, and Jun'ichi Tsujii. 2009. A Markov Logic Approach to Bio-Molecular Event Extraction. In *Proceedings of the BioNLP 2009 Workshop Companion Volume for Shared Task*, pages 41–49.

Laura Rimell and Stephen Clark. 2009. Porting a lexicalized-grammar parser to the biomedical domain. *Journal of Biomedical Informatics*, 42(5):852–865.

Jonathan Schuman and Sabine Bergler. 2006. Postnominal prepositional phrase attachment in proteomics. In *Proceedings of the Workshop on Linking Natural Language Processing and Biology*, pages 82–89.

Mihai Surdeanu, Richard Johansson, Adam Meyers, Lluís Màrquez, and Joakim Nivre. 2008. The CoNLL-2008 shared task on joint parsing of syntactic and semantic dependencies. In *CoNLL 2008: Proceedings of the Twelfth Conference on Natural Language Learning*, pages 159–177.

Andreas Vlachos, Paula Buttery, Diarmuid Ó Séaghdha, and Ted Briscoe. 2009. Biomedical event extraction without training data. In *Proceedings of the BioNLP 2009 Workshop Companion Volume for Shared Task*, pages 37–40.

Recognizing Biomedical Named Entities using Skip-chain Conditional Random Fields

Jingchen Liu **Minlie Huang*** **Xiaoyan Zhu**
Department of Computer Science and Technology
Tsinghua University, Beijing 100084, China
liu-jc04@mails.tsinghua.edu.cn
{aihuang, zxy-dcs}@tsinghua.edu.cn

Abstract

Linear-chain Conditional Random Fields (CRF) has been applied to perform the Named Entity Recognition (NER) task in many biomedical text mining and information extraction systems. However, the linear-chain CRF cannot capture long distance dependency, which is very common in the biomedical literature. In this paper, we propose a novel study of capturing such long distance dependency by defining two principles of constructing skip-edges for a skip-chain CRF: linking similar words and linking words having typed dependencies. The approach is applied to recognize gene/protein mentions in the literature. When tested on the BioCreAtIvE II Gene Mention dataset and GENIA corpus, the approach contributes significant improvements over the linear-chain CRF. We also present in-depth error analysis on inconsistent labeling and study the influence of the quality of skip edges on the labeling performance.

1 Introduction

Named Entity Recognition (NER) is a key task in most text mining and information extraction systems. The improvement in NER can benefit the final system performance. NER is a challenging task, particularly in the biomedical literature due to the variety of biomedical terminologies and the complicated syntactic structures.

Many studies have been devoted to biomedical NER. To evaluate biomedical NER systems, several challenge competitions had been held, such as BioNLP/NLPBA in 2004[1], BioCreAtIvE I in 2004 and BioCreAtIvE II in 2006[2]. The overview reports from these competitions, presenting state-of-the-art of biomedical NER studies, show that linear-chain Conditional Random Fields (CRF) is one of the most commonly used models and has the most competitive results (Yeh et al., 2005; Smith et al., 2008). Linear-chain CRF has also been successfully applied to other NLP tasks such as POS-tagging (Lafferty et al., 2001) and sentence chunking (Sha and Pereira, 2003). However, in most of these applications, only linear-chain CRF was fully exploited, assuming that only adjacent words are inter-dependent. The dependency between distant words, which occurs frequently in the biomedical literature, is yet to be captured.

In the biomedical literature, the repeated appearance of same or similar words in one sentence is a common type of long distance dependencies. This phenomenon is due to the complicated syntactic structures and the various biomedical terminologies in nature. See the following example:

> *"Both GH deficiency and impaired spinal growth may result in short stature, whereas the occurrence of early puberty in association with GH deficiency reduces the time available for GH therapy."*

the mentions of *GH* are repeated three times. If the entity are referred by a pronoun, the meaning of the sentence will be confusing and unclear because of the complex sentence structure. In this sentence:

> *"These 8-oxoguanine DNA glycosylases, hOgg1 (human) and mOgg1 (murine), are homologous to each other and to yeast Ogg1."*

the words *hOgg1*, *mOgg1* and *Ogg1* are homologous genes belonging to different species, having

* Corresponding author

[1] http://research.nii.ac.jp/~collier/workshops/JNLPBA04st.htm

[2] http://www.biocreative.org/

Proceedings of the 2010 Workshop on Biomedical Natural Language Processing, ACL 2010, pages 10–18,
Uppsala, Sweden, 15 July 2010.

very similar entity names. Some other types of long distance dependencies also occur frequently in the biomedical literature. For example, in this sentence

> *"Western immunoblot analysis detected p55gag and its cleavage products p39 and p27 in purified particles derived by expression of gag and gag-pol, respectively."*

the words *p55gag*, *p39* and *p27* conjuncted by *and*, have similar semantic meanings but they are separated by several tokens. A human curator can easily recognize such long distance dependencies and annotate these words consistently. However, when applying the linear-chain CRF, inconsistency errors in annotating these entities could happen due to the inability of representing long distance dependency.

In this paper, we present an approach of capturing long distance dependencies between words. We adopte the skip-chain CRF to improve the performance of gene mention recognition. We define two principles of connecting skip-edges for skip-chain CRF to capture long distance dependencies. The efficacy of the principles is investigated with extensive experiments. We test our method on two data sets and significant improvements are observed over the linear-chain CRF. We present in-depth error analysis on inconsistent labeling. We also investigat whether the quality of connected edges affect the labeling performance.

The remainder of this paper is organized as follows: We survey related studies in Section 2. We introduce linear-chain CRF and skip-chain CRF in Section 3. The method of connecting skip-chain edges is described in Section 4 . In Section 5 we present our experiments and in-depth analysis. We summarize our work in Section 6.

2 Related work

NER is a widely studied topic in text mining research, and many new challenges are seen in domain-specific applications, such as biomedical NER (Zhou et al., 2004). The dictionary based method is a common technique as biomedical thesauruses play a key role in understanding such text. Most dictionary based NER systems focused on: (1) integrating and normalizing different biomedical databases to improve the quality of the dictionary to be used; (2) improving matching strategies that are more suitable for biomedical terminologies; and (3) making filtering rules for post-processing to refine the matching results or to adjust the boundary of entities, see (Fukuda et al., 1998; Narayanaswamy et al., 2003; Yang et al., 2008). Many information extraction systems had a dictionary matching module to perform preliminary detection of named entities (Schuhmann et al., 2007; Kolarik et al., 2007; Wang et al., 2010).

Applying machine learning techniques generally obtains superior performance for the biomedical NER task. The automated learning process can induce patterns for recognizing biomedical names and rules for pre- and post-processing. Generally speaking, there are two categories of machine learning based methods: one treats NER as a classification task, while the other treats NER as a sequence labeling task. For the first category, Support Vector Machine (SVM) was a commonly adopted model (Kazama et al., 2002; Zhou et al., 2004). Lee et al. (2004) proposed a two-step framework to perform biomedical NER using SVM: firstly detecting the boundaries of named entities using classifiers; secondly classifying each named entity into predefined target types. For the second category, a sentence was treated as a sequence of tokens and the objective was to find the optimal label sequence for these tokens. The label space was often defined as {B,I,O}, where B indicates the beginning token of an entity, I denotes the continuing token and O represents the token outside an entity. The sequence labeling task can be approached by Hidden Markov Model (HMM), Conditional Random Field (CRF) , or a combination of different models (Zhou et al., 2005; Tatar and Cicekli, 2009).

Since proposed in (Lafferty et al., 2001), CRF has been applied to many sequence labeling tasks, including recognizing gene mentions from biomedical text (McDonald and Pereira, 2005). The Gene Mention Recognition task was included in both BioCreAtIvE I and BioCreAtIvE II challenges. CRF had been used in most of top performing systems in the Gene Mention Recognition task of BioCreAtIvE II (Smith et al., 2008). Some novel use of linear-chain CRF was proposed. For example, in (Kuo et al., 2007) labeling was performed in forward and backward directions on the same sentence and results were combined from the two directions. Huang et al. (2007) combines a linear-chain CRF and two SVM models to enhance the recall. Finkel et al. (2005) used Gibbs Sampling to add non-local dependencies into linear-chain CRF model for information extraction. However, the CRF models used in these systems were all linear-chain CRFs. To the best of our knowledge, no previous work has been done on using non-linear-chain CRF in the biomedical NER task.

Beyond the biomedical domain, skip-chain CRF has been used in several studies to model long distance dependency. In (Galley, 2006), skip edges were linked between sentences with non-local pragmatic dependencies to rank meetings. In (Ding et al., 2008), skip-chain CRF was used to detect the context and answers from online forums. The most close work to ours was in (Sutton and McCallum, 2004), which used skip-chain CRF to extract information from email messages announcing seminars. By linking the same words whose initial letter is capital, the method obtained improvements on extracting speakers' name. Our work is in the spirit of this idea, but we approach it in a different way. We found that the problem is much more difficult in the biomedical NER task: that is why we systematically studied the principles of linking skip edges and the quality of connected edges.

3 linear-chain and skip-chain CRF

Conditional Random Field is a probabilistic graphic model. The model predicts the output variables $\mathbf{y}$ for each input variables in $\mathbf{x}$ by calculating the conditional probability $p(\mathbf{y}|\mathbf{x})$ according to the graph structure that represents the dependencies between the $\mathbf{y}$ variables. Formally, given a graph structure over $\mathbf{y}$, the CRF model can be written as:

$$p(\mathbf{y}|\mathbf{x}) = \frac{1}{Z(\mathbf{x})} \prod_{C_p \in \zeta} \prod_{\Psi_c \in C_p} \Psi_c(\mathbf{x}_c, \mathbf{y}_c; \theta_p) \quad (1)$$

$Z(\mathbf{x})$ is a normalization factor.

In this definition, the graph is partitioned into a set of cliques $\zeta = \{C_1, C_2, \ldots C_p\}$, where each C_p is a clique template. Each Ψ_c, called a factor, is corresponding to one edge in the clique c, and can be parameterized as:

$$\Psi_c(\mathbf{x}_c, \mathbf{y}_c; \theta_p) = \exp \sum_{k=1} \lambda_{pk} f_{pk}(\mathbf{x}_c, \mathbf{y}_c) \quad (2)$$

Each feature function $f_{pk}(\mathbf{x}_c, \mathbf{y}_c)$ represents one feature of $\mathbf{x}$ and the λ_{pk} is the feature weight.

In the training phrase, the parameters is estimated using an optimization algorithm such as limited memory BFGS etc. In the testing phrase, CRF finds the most likely label sequence for an unseen instance by maximizing the probability defined in (1).

In the NER task, one sentence is firstly tokenized into a sequences of tokens and each token can be seen as one word. Each node in the graph is usually corresponding to one word in a sentence. Each x variable represents a set of features for one word, and each y is the variable for the label of one word. Note that when one edge is linked between two words, the edge is actually linked between their corresponding y variables. The $\mathbf{y}$ label is one of {B,I,O}, in which B means the beginning word of an entity, I means the inside word of an entity, and O means outside an entity.

If we link each word with its immediate preceding words to form a linear structure for one sentence, we get a linear-chain CRF, defined as:

$$p_\theta(\mathbf{y}|\mathbf{x}) = \frac{1}{Z(\mathbf{x})} \prod_{t=1}^{T} \Psi_t(y_t, y_{t-1}, \mathbf{x}) \quad (3)$$

This structure contains only one clique template. If we add an extra clique template that contains some skip edges between nonadjacent words, the CRF become a skip-chain CRF, formulated as follows:

$$p_\theta(\mathbf{y}|\mathbf{x}) = \frac{1}{Z(\mathbf{x})} \prod_{t=1}^{T} \Psi_t(y_t, y_{t-1}, \mathbf{x}) \cdot \prod_{(u,v) \in \tau} \Psi_{uv}(y_u, y_v, \mathbf{x}) \quad (4)$$

τ is the edge set of the extra clique template containing skip edges. An illustration of linear-chain and skip-chain CRF is given in Figure 1. It is straightforward to change a linear-chain CRF to a skip-chain CRF by simply linking some additional skip edges. However, it must be careful to add such edges because different graph structures require different inference algorithms. Those inference algorithms may have quite different time complexity. For example, for the linear-chain CRF, inference can be performed efficiently and exactly by a dynamic-programming algorithm. However, for the non-linear structure, approximate inference algorithms must be used. Solving arbitrary CRF graph structures is NP-hard. In other word, we must be careful to link too many

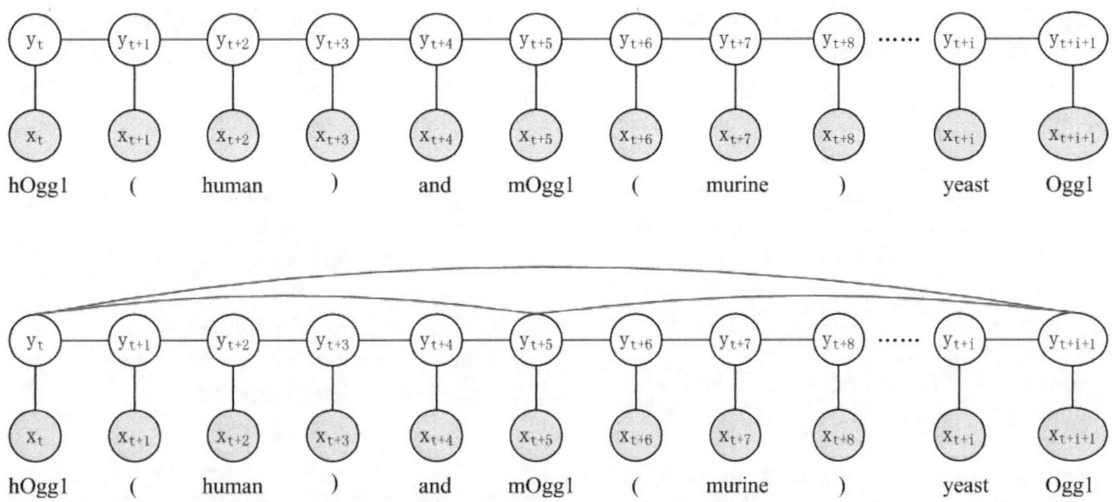

Figure 1: The illustration of linear-chain CRF and skip-chain CRF. The blue edges represent the linear-chain edges belonging to one clique template, while the red edges represent the skip edges belonging to another clique template.

skip edges to avoid making the model impractical. Therefore, it is absolutely necessary to study which kinds of edges will contribute to the performance while avoiding over-connected edges.

3.1 Features

As our interest is in modifying the CRF graph structure rather than evaluating the effectiveness of features, we simply adopted features from the state-of-the-art such as (McDonald and Pereira, 2005) and (Kuo et al., 2007).

- **Common Features**: the original word, the stemmed word, the POS-tag of a word, the word length, is or not the beginning or ending word of the sentence etc.
- **Regular Expression Features**: a set of regular expressions to extract orthographic features for the word.
- **Dictionary Features**: We use several lexicons. For example, a protein name dictionary compiled from SWISS-PROT, a species dictionary from NCBI Taxonomy, a drug name dictionary from DrugBank database, and a disease name dictionary from several Internet web site.
- **N-gram Features**: For each token, we extract the corresponding 2-4 grams into the feature set.

Each word will include the adjacent words' features within $\{-2, -1, 0, 1, 2\}$ offsets. The features used in the linear-chain CRF and skip-chain CRF are all the same in our experiment.

4 Method

As the limitations discussed above, detecting the necessary nodes to link should be the first step in constructing a skip-chain CRF. In the speaker name extraction task (Sutton and McCallum, 2004), only identical capitalized words are linked, because there is few variations in the speaker's name. However, gene mentions often involve words without obvious orthographic features and such phenomena are common in the biomedical literature such as *RGC DNA sequence* and *multisubunit TFIID protein*. If we link all the words like *DNA*, *sequence* and *protein*, the efficiency and performance will drop due to over-connected edges. Therefore, the most important step of detecting gene mentions is to determine which edges should be connected.

4.1 Detect keywords in gene mention

We found that many gene mentions have at least one important word for the identification of gene mentions. For example, the word, *Gal4*, is such a keyword in *Gal4 protein* and *NS1A* in *NS1A protein*. These words can distinguish gene mentions from other common English words and phrases, and can distinguish different gene mentions as well. We define such words as the keyword of a gene mention. The skip edges are limited to only connect these keywords. We use a rule-based method to detect keywords. By examining the annotated data, we defined keywords as those containing at least one capital letter or digit. And at the same time, keywords must conform to the following rules:

- Keywords are not stop words, single letters, numbers, Greek letters, Roman numbers or nucleotide sequence such as *ATTCCCTGG*.

- Keywords are not in the form of an uppercase initial letter followed by lowercase letters, such as *Comparison* and *Watson*. These words have capital letters only because they are the first word in the sentences, or they are the names of people or other objects. This rule will miss some correct candidates, but reduces noise.

- Keywords do not include some common words with capital letters such as *DNA*, *cDNA*, *RNA*, *mRNA*, *tRNA* etc. and some frequently appearing non-gene names such as *HIV* and *mmHg*. We defined a lexicon for such words on the training data.

4.2 Link similar keywords

After keyword candidates are detected, we judge each pair of keywords in the same sentence to find similar word pairs. Each word pair is examined by these rules:

- They are exactly the same words.

- Words only differ in digit letters, such as *CYP1* and *CYP2*.

- Words with the same prefix, such as *IgA* and *IgG*, or with the same suffix, such as *ANF* and *pANF*.

The token pair will be linked by a skip edge if they match at least one rule.

4.3 Link typed dependencies

Some long distance dependency cannot be detected simply by string similarity. To capture such dependency, we used stanford parser[3] to parse sentences and extract typed dependencies from parsed results. The typed dependencies are a set of binary relations belonging to 55 pre-defined types to provide a description of the grammatical relationships in a sentence (Marneffe and Manning, 2008). Some examples of typed dependencies are listed in Table 1.

Type	Description
conj	conjuncted by the conjunction such as *and*
prep	prepositional modifier
nn	noun compound modifier
amod	adjectival modifier
dep	uncertain types

Table 1: Examples for typed dependencies.

The output of the parser is pairs of dependent words, along with typed dependencies between two words in a pair. For example, in the sentence:

> *"...and activate transcription of a set of genes that includes G1 cyclins CLN1, CLN2, and many DN, synthesis genes."*

a typed dependency *nn(G1,CLN1)* is extracted by the parser, meaning the words *G1* and *CLN1* has a typed dependency of *nn* because they form a noun phrase under a dependency grammar: modification. Similarly, in the sentence

> *"Using the same approach we have shown that hFIRE binds the stimulatory proteins Sp1 and Sp3 in addition to CBF."*

the words *Sp1* and *Sp3* can be detected to have a typed dependency of *conj_and*, and the two words have a typed denpendency of *prep_in_addition_to* with *CBF*, respectively. The most common type dependencies are *conj_and*, *nn* and *dep*. The keywords having typed dependencies will be linked by a skip edge.

5 Experiment

We tested our method on two datasets: the Gene Mention (GM) data in BioCreAtIvE II (BCIIGM)

[3] `http://nlp.stanford.edu/software/lex-parser.shtml`

[4]and GENIA corpus[5]. The BCIIGM dataset was used in the BioCreAtIvE II Gene Mention Recognition task in 2006. It was built from the GENETAG corpus (Tanabe et al., 2005) with some modification of the annotation. The dataset contains 15000 sentences for training and 5000 sentences for testing. Two gold-standard sets, GENE and ALTGENE, were provided for evaluation and an official evaluation procedure in Perl script was provided. The ALTGENE set provides alternate forms for genes in the GENE set. In the official evaluation, each identified string will be looked up in both GENE and ALTGENE. If the corresponding gene was found in either GENE or ALTGENE, the identified string will be counted as a correct answer.

The GENIA corpus is a widely used dataset in many NER and information extraction tasks due to its high quality annotation. The GENIA corpus contains 2000 abstracts from MEDLINE, with approximately 18500 sentences. The corpus was annotated by biomedical experts according to a predefined GENIA ontology. In this work, we only used the annotated entities that have a category of protein, DNA, or RNA. These categories are related to the definition of gene mention in BioCreAtIvE II. We only used strict matching evaluation (no alternate forms check) for the GENIA corpus as no ALTGENE-like annotation is available.

The performance is measured by precision, recall and F score. Each identified string is counted as a true positive (TP) if it is matched by a gold-standard gene mention, otherwise the identified string is a false positive (FP). Each gold standard gene mention is counted as a false negative (FN) if it is not identified by the approach. Then the precision, recall and their harmonic average F score is calculated as follows:

$$precision = \frac{TP}{TP + FP}$$

$$recall = \frac{TP}{TP + FN}$$

$$F = \frac{2 \cdot precision \cdot recall}{precision + recall}$$

To implement both linear-chain CRF and skip-chain CRF, we used the GRMM Java package[6] which is an extended version of MALLET. The package provides an implement of arbitrary structure CRF.

5.1 Result Comparison

We evaluated our approach on the BCIIGM dataset and GENIA corpus. For the BCIIGM dataset, two evaluation criteria were used: *official* - exactly the same as that used in the BioCreAtIvE II competition, with the official evaluation procedure; and *strict* - strict matching for each identified string without checking its alternate forms in ALTGENE. The GENIA dataset were randomly divided into 10 parts to perform a 10-fold cross validation. However, we didn't do cross validation on the BCIIGM dataset because the BioCreAtIvE II competition annotations and evaluation procedure were tailored to evaluating participating systems.

The comparative results are listed in Table 2. We compared the two edge linking principles, linking similar words and linking words having typed dependencies. The F score from the skip-chain CRF is better than that from the linear-chain CRF. Significance tests were performed to check whether these results have significant differences. Paired two-tail t-tests were conducted with respect to the F scores of linear-chain CRF vs. those of the two skip-chain CRFs, respectively. The p-value was 1.989×10^{-7} for the skip-chain CRF linked by similar words vs. linear-chain CRF. The p-value was 3.971×10^{-5} for the skip-chain CRF linked by typed dependencies vs. linear-chain CRF. This shows that the improvement is significant.

Note that we did not compare our results on the BCIIGM dataset to those submitted to the competition. There are two reasons for this: First, our focus is on comparing the skip-chain CRF with the linear-chain CRF. Second, in the competition, most participating systems that used CRF also applied other algorithms, or sophisticated rules for adjusting detected boundaries or refining the recognized results, to achieve competitive performance. By contrast, we did not employ any post-processing rule or algorithm to further improve the performance. In this sense, comparing our results to those has become unfair.

[4]http://sourceforge.net/projects/biocreative/files/

[5]http://www-tsujii.is.s.u-tokyo.ac.jp/GENIA/home/wiki.cgi?page=Technical+Term+Annotation

[6]http://mallet.cs.umass.edu/grmm/index.php

Data	Model	Precision(%)	Recall(%)	F score(%)
BCIIGM official	linear-chain CRF	85.16	81.50	83.29
	skip-chain CRF linked by sim-words	86.68	82.75	84.67
	skip-chain CRF linked by typed-dep	86.73	82.36	84.49
BCIIGM strict	linear-chain CRF	74.09	69.49	71.73
	skip-chain CRF linked by sim-words	76.26	71.53	73.82
	skip-chain CRF linked by typed-dep	75.99	70.49	73.14
GENIA	linear-chain CRF	76.77	74.92	75.83
	skip-chain CRF linked by sim-words	78.57	77.12	77.82
	skip-chain CRF linked by typed-dep	78.18	76.87	77.52

Table 2: The result comparison between the linear-chain CRF and skip-chain CRF. *BCIIGM* is the BioCreAtIvE II Gene Mention Recognition dataset. *official* means using the official provided evaluation procedure and *strict* means using strict matching to evaluate the results. *sim-words* means similar words and *typed-dep* means typed dependencies. The results for GENIA are averaged over 10-fold cross validation.

5.2 Discussion

We provided in-depth analysis of our results on the BCIIGM dataset. As one of our motivations for connecting words with skip edges is to enhance the consistency of labeling, we firstly examined whether the proposed approach can provide consistent labeling. Let us start from two typical examples. In the first sentence

> *"The response sequences were localized between -67 and +30 in the simian cytomegalovirus IE94 promoter and upstream of position +9 in the HCMV IE68 promoter."*

the word *IE94* is missed (not labeled) while its similar word *IE68* is labeled correctly by the linear-chain CRF. In the second sentence

> *"It is suggested that biliary secretion of both TBZ and FBZ and their metabolites may contribute to this recycling."*

the word *TBZ* is labeled as a gene mention incorrectly (false positive) while its similar word *FBZ* is not labeled at all (true negative) by the linear-chain CRF. Both sentences are correctly labeled by the skip-chain CRF. Similar improvements are also made by the skip-chain CRF model linked by typed dependencies. To study labeling consistency, we counted the statistics of inconsistency errors, as shown in Table 3. Two kinds of inconsistency errors were counted: false negatives correctable by consistency (FNCC) and false positives correctable by consistency (FPCC). An FNCC means that a gold-standard mention is missed by the system while its skip edge linked gene mention is correctly labeled, which is similar to the *inconsistent miss* in (Sutton and McCallum, 2004), as the *IE94* in the first example. An FPCC means a non-gene mention is labeled as a gene while its skip edge linked mention (also non-gene mention) is not recognized, as *TBZ* in the second example. These two kinds of inconsistency errors lead to inconsistent false negatives (FN) and false positives (FP). A good model should reduce as much inconsistency errors as possible. The inconsistency errors are reduced substantially as we expected, showing that the reduction of inconsistency errors is one reason for the performance improvements.

The skip-chain CRF linked by similar words had better performance than the skip-chain CRF linked by typed dependencies. This may infer that the quality of skip edges has impact on the performance. In order to study this issue, the quality of skip edges was examined. The statistics of skip edges in the BCIIGM dataset for the two skip-chain CRF models (linked by similar words and by typed dependencies respectively) is shown in the first two rows of Table 4. A skip edge is counted as a correct edge if the edge links two words that are both gene mentions in the gold-standard annotation. The statistics shows that the skip-chain CRF linked by similar words has a higher precision than the model by typed dependencies. To make the comparison more evident, we built another skip-chain CRF whose skip edges were randomly connected. The number of skip edges in this model

Skip edge type	Model	FPCC	FNCC
sim-words	linear-chain	112	70
	skip-chain	48	20
Percentage of reduction		57.14%	71.43%
typed-dep	linear-chain	32	29
	skip-chain	9	5
Percentage of reduction		71.88%	82.76%

Table 3: Statistics of inconsistency errors for the linear-chain CRF and skip-chain CRF. *FPCC* is false positives correctable by consistency and *FNCC* is false negatives correctable by consistency in the table. The percentage is calculated by dividing the reduction of errors by the error number of linear-chain CRF, for example $(112 - 48)/48 = 57.14\%$.

approximately equals to that in the skip-chain CRF linked by similar words. The percentage of correct skip-edges in this model is small, as shown in the last row of Table 4. We tested this skip-chain CRF model on the BCIIGM dataset under the strict matching criterion. The performance of the randomly linked skip-chain CRF is shown in Table 5. As can be seen from the table, the performance of the randomly connected skip-chain CRF droped remarkably, even worse than that of the linear-chain CRF. This confirms that the quality of skip edges is a key factor for the performance improvement.

Model	Edges	Correct edges	Percentage
sim-words	1912	1344	70.29%
typed-dep	728	425	53.38%
random	1906	41	2.15%

Table 4: Statistics of skip edges and correct skip edges for the skip-chain CRF models. *sim-words* means the skip-chain CRF linked by similar words, *typed-dep* means the CRF linked by typed dependencies and *random* means the skip-chain CRF has randomly connected skip edges. The edges are counted in the BCIIGM testing data.

From the above discussion, we summarize this section as follows: (1) the skip-chain CRF with high quality skip edges can reduce inconsistent labeling errors, and (2) the quality of skip edges is crucial to the performance improvement.

Model	P (%)	R(%)	F(%)
linear	74.09	69.49	71.73
sim-words	76.26	71.53	73.82
typed-dep	75.99	70.49	73.14
random	73.66	69.13	71.32

Table 5: Performance comparison between the randomly linked skip-chain CRF and other models. The result was tested on the BCIIGM dataset under the strict matching criterion. *P*, *R* and *F* denote the precision, recall and F score respectively. *linear* denotes the linear-chain CRF. *sim-words* denotes the skip-chain CRF linked by similar words. *typed-dep* denotes the skip-chain CRF linked by typed dependencies. *random* denotes the skip-chain CRF having randomly linked skip edges.

6 Conclusion

This paper proposed a method to construct a skip-chain CRF to perform named entity recognition in the biomedical literature. We presented two principles to connect skip edges to address the issue of capturing long distance dependency: linking similar keywords and linking words having typed dependencies. We evaluated our method on the BioCreAtIvE II GM dataset and GENIA corpus. Significant improvements were observed. Moreover, we presented in-depth analysis on inconsistent labeling errors and the quality of skip edges. The study shows that the quality of linked edges is a key factor of the system performance.

The quality of linked edges plays an important role in not only performance but also time efficiency. Thus, we are planning to apply machine learning techniques to automatically induce patterns for linking high-quality skip-edges. Furthermore, to refine the recognition results, we are planning to employ post-processing algorithms or construct refinement rules.

Acknowledgments

This work was partly supported by the Chinese Natural Science Foundation under grant No. 60803075 and No.60973104, and partly carried out with the aid of a grant from the International Development Research Center, Ottawa, Canada IRCI project from the International Development.

References

Shilin Ding, Gao Cong, Chin-Yew Lin and Xiaoyan Zhu. 2008. *Using Conditional Random Fields to Extract Contexts and Answers of Questions from Online Forums*. In Proceedings of 46th Annual Meeting of the Association for Computational Linguistics (ACL'08), pp 710-718.

Jenny Rose Finkel, Trond Grenager, and Christopher Manning. 2005. *Incorporating Non-local Information into Information Extraction Systems by Gibbs Sampling*. Proceedings of the 43rd Annual Meeting of the ACL, pages 363C370.

K. Fukuda, A. Tamura, T. Tsunoda and T. Takagi. 1998. *Toward information extraction: identifying protein names from biological papers*. Pacific Symposium on Biocomputing. 1998.

Michel Galley. 2006. *A Skip-Chain Conditional Random Field for Ranking Meeting Utterances by Importance*. Proceedings of the 2006 Conference on Empirical Methods in Natural Language Processing (EMNLP 2006), pages 364-372.

Han-Shen Huang, Yu-Shi Lin, Kuan-Ting Lin, Cheng-Ju Kuo, Yu-Ming Chang, Bo-Hou Yang, I-Fang Chung and Chun-Nan Hsu. 2007. *High-recall gene mention recognition by unification of multiple backward parsing models*. Proceedings of the Second BioCreative Challenge Evaluation Workshop, pages 109-111.

Jun'ichi Kazama, Takaki Makino, Yoshihiro Ohta and Jun'ichi Tsujii. 2002. *Tuning support vector machines for biomedical named entity recognition*. Proceedings of the ACL-02 workshop on Natural language processing in the biomedical domain - Volume 3.

Corinna Kolarik, Martin Hofmann-Apitius, Marc Zimmermann and Juliane Fluck. 2007. *Identification of new drug classification terms in textual resources*. Bioinformatics 2007 23(13):i264-i272

Cheng-Ju Kuo, Yu-Ming Chang, Han-Shen Huang, Kuan-Ting Lin, Bo-Hou Yang, Yu-Shi Lin, Chun-Nan Hsu and I-Fang Chung. 2007. *Rich Feature Set, Unification of Bidirectional Parsing and Dictionary Filtering for High F-Score Gene Mention Tagging*. Proceedings of the Second BioCreative Challenge Evaluation Workshop, pages 105-107.

Ki-Joong Lee, Young-Sook Hwang, Seonho Kim and Hae-Chang Rim. 2004. *Biomedical named entity recognition using two-phase model based on SVMs*. Journal of Biomedical Informatics, Volume 37, Issue 6, December 2004, Pages 436-447.

John Lafferty, Andrew McCallum and Fernando Pereira. 2001. *Conditional random fields: Probabilistic models for segmenting and labeling sequence data*. In Proc. ICML-01, pages 282-289, 2001.

Ryan McDonald and Fernando Pereira. 2005. *Identifying gene and protein mentions in text using conditional random fields*. BMC Bioinformatics 2005, 6(Suppl 1):S6.

Marie-Catherine de Marneffe and Christopher D. Manning. 2008. *Stanford typed dependencies manual*.

M. Narayanaswamy, K.E. Ravikumar and K. Vijay-Shanker. 2003. *A biological named entity recognizer*. Pacific Symposium on Biocomputing. 2003.

Dietrich Rebholz-Schuhmann, Harald Kirsch, Miguel Arregui, Sylvain Gaudan, Mark Riethoven and Peter Stoehr. 2007. *EBIMedltext crunching to gather facts for proteins from Medline*. Bioinformatics 2007 23(2):e237-e244

Fei Sha and Fernando Pereira. 2003. *Shallow Parsing with Conditional Random Fields*. Proceedings of HLT-NAACL 2003, Main Papers, pp.134-141

Larry Smith, Lorraine K Tanabe, et al. 2008. *Overview of BioCreative II gene mention recognition*. Genome Biology 2008, 9(Suppl 2):S2.

Charles Sutton and Andrew McCallum. 2004. *Collective Segmentation and Labeling of Distant Entities in Information Extraction*. ICML workshop on Statistical Relational Learning, 2004.

Lorraine Tanabe, Natalie Xie, Lynne H Thom, Wayne Matten and W John Wilbur. 2005. *GENETAG: a tagged corpus for gene/protein named entity recognition* . BMC Bioinformatics 2005, 6(Suppl 1):S3

Serhan Tatar and Ilyas Cicekli. 2009. *Two learning approaches for protein name extraction*. Journal of Biomedical Informatics 42(2009) 1046-1055

Xinglong Wang, Jun'ichi Tsujii and Sophia Ananiadou. 2010. *Disambiguating the species of biomedical named entities using natural language parsers*. Bioinformatics 2010 26(5):661-667

Zhihao Yang, Hongfei Lin and Yanpeng Li. 2008. *Exploiting the performance of dictionary-based bioentity name recognition in biomedical literature*. Computational Biology and Chemistry 32(2008) 287-291.

Alexander Yeh, Alexander Morgan, Marc Colosimo and Lynette Hirschman. 2005. *BioCreAtIvE Task 1A: gene mention finding evaluation*. BMC Bioinformatics 2005, 6(Suppl 1):S2.

GuoDong Zhou, Jie Zhang, Jian Su, Dan Shen, ChewLim Tan. 2004. *Recognizing names in biomedical texts: a machine learning approach*. Bioinformatics 2004, Vol.20(7),pp.1178C1190.

GuoDong Zhou, Dan Shen, Jie Zhang, Jian Su1 and SoonHeng Tan. 2005. *Recognition of protein/gene names from text using an ensemble of classifiers*. BMC Bioinformatics 2005, 6(Suppl 1):S7.

Event Extraction for Post-Translational Modifications

Tomoko Ohta* **Sampo Pyysalo*** **Makoto Miwa*** **Jin-Dong Kim*** **Jun'ichi Tsujii*†‡**
*Department of Computer Science, University of Tokyo, Tokyo, Japan
†School of Computer Science, University of Manchester, Manchester, UK
‡National Centre for Text Mining, University of Manchester, Manchester, UK
{okap,smp,mmiwa,jdkim,tsujii}@is.s.u-tokyo.ac.jp

Abstract

We consider the task of automatically extracting post-translational modification events from biomedical scientific publications. Building on the success of event extraction for phosphorylation events in the BioNLP'09 shared task, we extend the event annotation approach to four major new post-transitional modification event types. We present a new targeted corpus of 157 PubMed abstracts annotated for over 1000 proteins and 400 post-translational modification events identifying the modified proteins and sites. Experiments with a state-of-the-art event extraction system show that the events can be extracted with 52% precision and 36% recall (42% F-score), suggesting remaining challenges in the extraction of the events. The annotated corpus is freely available in the BioNLP'09 shared task format at the GENIA project homepage.[1]

1 Introduction

Post-translational-modifications (PTM), amino acid modifications of proteins after translation, are one of the posterior processes of protein biosynthesis for many proteins, and they are critical for determining protein function such as its activity state, localization, turnover and interactions with other biomolecules (Mann and Jensen, 2003). Since PTM alter the properties of a protein by attaching one or more biochemical functional groups to amino acids, understanding of the mechanism and effects of PTM are a major goal in the recent molecular biology, biomedicine and pharmacology fields. In particular, epigenetic ("outside conventional genetics") regulation of gene expression has a crucial role in these fields and PTM-like modifications of biomolecules are a burning issue. For instance, tissue specific or context dependent expression of many proteins is now known to be controlled by specific PTM of histone proteins, such as *Methylation* and *Acetylation* (Jaenisch and Bird, 2003). This *Methylation* and *Acetylation* of specific amino acid residues in histone proteins are strongly implicated in unwinding the nucleosomes and exposing genes to transcription, replication and DNA repairing machinery.

The recent BioNLP'09 Shared Task on Event Extraction (Kim et al., 2009a) (below, BioNLP shared task) represented the first community-wide step toward the extraction of fine-grained event representations of information from biomolecular domain publications (Ananiadou et al., 2010). The nine event types targeted in the task included one PTM type, *Phosphorylation*, whose extraction involved identifying the modified protein and, when stated, the specific phosphorylated site. The results of the shared task showed this PTM event to be single most reliably extracted event type in the data, with the best-performing system for the event type achieving 91% precision and 76% recall (83% F-score) in the extraction of phosphorylation events (Buyko et al., 2009). The results suggest both that the event representation is well applicable to PTM and that current extraction methods are capable of reliable PTM extraction. Most of the proposed state-of-the-art methods for event extraction are further largely machine-learning based. This suggest that the coverage of many existing methods could be straightforwardly extended to new event types and domains by extending the scope of available PTM annotations and retraining the methods on newly annotated data. In this study, we take such an annotation-based approach to extend the extraction capabilities of state of the art event extraction methods for PTM.

[1] http://www-tsujii.is.s.u-tokyo.ac.jp/GENIA

Proceedings of the 2010 Workshop on Biomedical Natural Language Processing, ACL 2010, pages 19–27,
Uppsala, Sweden, 15 July 2010.

Term	Count	
Phosphorylation	172875	50.90%
Methylation	49780	14.66%
Glycosylation	36407	10.72%
Hydroxylation	20141	5.93%
Acetylation	18726	5.51%
Esterification	7836	2.31%
Ubiquitination	6747	1.99%
ADP-ribosylation	5259	1.55%
Biotinylation	4369	1.29%
Sulfation	3722	1.10%
...		
TOTAL	339646	100%

Table 1: PTM mentions in PubMed. The number of citations returned by the PubMed search engine for each PTM term shown together with the fraction of the total returned for all searches. Searches were performed with the terms as shown, allowing MeSH term expansion and other optimizations provided by the Entrez search.

2 Corpus Annotation

We next discuss the selection of the annotated PTM types and source texts and present the representation and criteria used in annotation.

2.1 Event Types

A central challenge in the automatic extraction of PTMs following the relatively data-intensive BioNLP shared task model is the sheer number of different modifications: the number of known PTM types is as high as 300 and constantly growing (Witze et al., 2007). Clearly, the creation of a manually annotated resource with even modest coverage of statements of each of the types would be a formidable undertaking. We next present an analysis of PTM statement occurrences in PubMed as the first step toward resolving this challenge.

We estimated the frequency of mentions of prominent PTM types by combining MeSH ontology[2] PTM terms with terms occurring in the `post-translational protein modification` branch of the Gene Ontology (The Gene Ontology Consortium, 2000). After removing variants (e.g. *polyamination* for *amination* or *dephosphorylation* for *phosphorylation*) and two cases judged likely to occur frequently in non-PTM contexts (*hydration* and *oxidation*), we searched PubMed for the remaining 31 PTM types. The results for the most frequent types are shown in Table 1. We find a power-law - like distribution with *phosphorylation* alone accounting for over 50% of the total, and the top 6 types together for over 90%. By contrast, the bottom ten types together represent less than a percent of total occurrences.

This result implies that fair coverage of individual PTM event mentions can be achieved without considering even dozens of different PTM event types, let alone hundreds. Thus, as a step toward extending the coverage of event extraction systems for PTM, we chose to focus limited resources on annotating a small selection of types so that a number of annotations sufficient for supervised learning and stable evaluation can be provided. To maximize the utility of the created annotation, the types were selected based on their frequency of occurrence.

2.2 Text Selection

Biomedical domain corpora are frequently annotated from selections of texts chosen as a sample of publications in a particular subdomain of interest. While several areas in present-day molecular biology are likely to provide ample source data for PTM statements, a sample of articles from any subdomain is unlikely to provide a well-balanced distribution of event types: for example, the most frequent PTM event type annotated in the GENIA event corpus occurs more than 10 times as often as the second most frequent (Kim et al., 2008). Further, avoiding explicit subdomain restrictions is not alone sufficient to assure a balanced distribution of event types: in the BioInfer corpus, for which sentences were selected on the basis of their containing mentions of protein pairs known to interact, the most frequent PTM type is again annotated nearly four times as often as the second most frequent (Pyysalo et al., 2007).

To focus annotation efforts on texts relevant to PTM and to guarantee that the annotation results in relatively balanced numbers of PTM events of each targeted type, we decided to annotate a targeted set of source texts instead of a random sample of texts for a particular subdomain. This type of targeted annotation involves a risk of introducing bias: a badly performed selection could produce a corpus that is not representative of the

[2] `http://www.nlm.nih.gov/mesh/meshhome.html`

PTM type	AB	FT
Acetylation	103	128
Glycosylation	226	336
Methylation	72	69
Phosphorylation	186	76
Hydroxylation	71	133

Table 2: Number of abstracts (AB) and full-text articles (FT) tagged in PIR as containing PTM statements.

statements expressing PTMs in text and thus poor material for either meaningful evaluation or for training methods with good generalization performance.[3] To avoid such bias, we decided to base our selection of the source texts on an independently annotated PTM resource with biological (as opposed to textual) criteria for inclusion. Owing in part to the recent interest in PTMs, there are currently a wealth of resources providing different levels of annotation for PTMs.

Here, we have chosen to base initial annotation on corpora provided by the Protein Information Resource[4] (PIR) (Wu et al., 2003). These corpora contain annotation for spans with evidence for five different PTM types (Table 2), corresponding to the five PTMs found above to occur in PubMed with the highest frequency. A key feature setting this resource apart from others we are aware of is that it provides text-bound annotations identifying the statement by which a PTM record was made in the context of the full publication abstracts. While this annotation is less specific and detailed than the full BioNLP shared task markup, it could both serve as an initial seed for annotation and assure that the annotation agrees with relevant database curation criteria. The PIR corpora have also been applied in previous PTM extraction studies (e.g. (Hu et al., 2005; Narayanaswamy et al., 2005)).

We judged that the annotated Phosphorylation events in the BioNLP shared task data provide sufficient coverage for the extraction of this PTM type, and chose to focus on producing annotation for the four other PTM types in the PIR data. As the high extraction performance for phosphorylation events in the BioNLP shared task was

Protein	Site	PTM	Count
collagen	lysine	Hydroxylate	44
myelin	arginine	Methylate	17
M protein	N-terminal	Glycosylate	2
EF-Tu	lysine	Methylate	1
Actobindin	NH2 terminus	Acetylate	0

Table 3: Example queried triples and match counts from Medie.

achieved with annotated training data containing 215 PTM events, in view of the available resources we set as an initial goal the annotation of 100 events of each of the four PTM types. To assure that the annotated resource can be made publicly available, we chose to use only the part of the PIR annotations that identified sections of PubMed abstracts, excluding full-text references and non-PubMed abstracts. Together with the elimination of duplicates and entries judged to fall outside of the event annotation criteria (see Section 2.4), this reduced the number of source texts below our target, necessitating a further selection strategy.

For further annotation, we aimed to select abstracts that contain specific PTM statements identifying both the name of a modified protein and the modified site. As for the initial selection, we further wished to avoid limiting the search by searching for any specific PTM expressions. To implement this selection, we used the Medie system[5] (Ohta et al., 2006; Miyao et al., 2006) to search PubMed for sentences where a specific protein and a known modified site were found together in a sentence occurring in an abstract annotated with a specific MeSH term. The (protein name, modified site, MeSH term) triples were extracted from PIR records, substituting the appropriate MeSH term for each PTM type. Some examples with the number of matching documents are shown in Table 3. As most queries returned either no documents or a small number of hits, we gave priority to responses to queries that returned a small number of documents to avoid biasing the corpus toward proteins whose modifications are frequently discussed.

We note that while the PIR annotations typically identified focused text spans considerably shorter than a single sentence and sentence-level search was used in the Medie-based search to increase the likelihood of identifying relevant statements, after selection all annotation was performed to full abstracts.

[3]One could easily gather PTM-rich texts by performing protein name tagging and searching for known patterns such as "[PROTEIN] methylates [PROTEIN]", but a corpus created in this way would not necessarily provide significant novelty over the original search patterns.

[4]`http://pir.georgetown.edu`

[5]`http://www-tsujii.is.s.u-tokyo.ac.jp/medie/`

Event type	Count
Protein_modification	38
Phosphorylation	546
Dephosphorylation	28
Acetylation	7
Deacetylation	1
Ubiquitination	6
Deubiquitination	0

Table 4: GENIA PTM-related event types and number of events in the GENIA event corpus. Type names are simplified: the full form of e.g. the *Phosphorylation* type in the GENIA event ontology is *Protein_amino_acid_phosphorylation*.

Event type	Arguments	Count
Protein_modification	Theme	31
Phosphorylation	Theme	261
Phosphorylation	Theme, Site	230
Phosphorylation	Site	20
Phosphorylation	Theme, Cause	14
Dephosphorylation	Theme	16

Table 5: GENIA PTM-related event arguments. Only argument combinations appearing more than 10 times in the corpus shown.

2.3 Representation

The employed event representation can capture the association of varying numbers of participants in different roles. To apply an event extraction approach to PTM, we must first define the targeted representation, specifying the event types, the mandatory and optional arguments, and the argument types – the roles that the participants play in the events. In the following, we discuss alternatives and present the representation applied in this work.

The GENIA Event ontology, applied in the annotation of the GENIA Event corpus (Kim et al., 2008) that served as the basis of the BioNLP shared task data, defines a general *Protein_modification* event type and six more specific modification subtypes, shown in Table 4. While the existing *Acetylation* type could thus be applied together with the generic *Protein_modification* type to capture all the annotated PTMs, we believe that identification of the specific PTM type is not only important to users of extracted PTM events but also a relatively modest additional burden for automatic extraction, owing to the unambiguous nature of typical expressions used to state

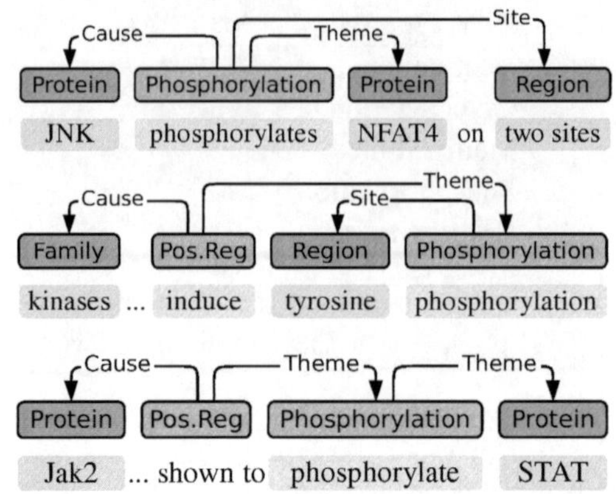

Figure 1: Alternative representations for PTM statements including a catalyst in GENIA Event corpus. PTM events can be annotated with a direct Cause argument (top, PMID 9374467) or using an additional *Regulation* event (middle, PMID 10074432). The latter annotation can be applied also in cases where there is no expression directly "triggering" the secondary event (bottom, PMID 7613138).

PTMs in text. We thus chose to introduce three additional specific modification types, *Glycosylation*, *Hydroxylation* and *Methylation* for use in the annotation.

The GENIA Event corpus annotation allows PTM events to take Theme, Site and Cause arguments specifying the event participants, where the Theme identifies the entity undergoing the modification, Site the specific region being modified, and Cause an entity or event leading to the modification. Table 5 shows frequent argument combinations appearing in the annotated data. We note that while Theme is specified in the great majority of events and Site in almost half, Cause is annotated for less than 5% of the events. However, the relative sparsity of Cause arguments in modification events does not imply that e.g. catalysts of the events are stated only very rarely, but instead reflects also the use of an alternative representation for capturing such statements without a Cause argument for the PTM event. The GENIA event annotation specifies a *Regulation* event (with *Positive_regulation* and *Negative_regulation* subtypes), used to annotate not only regulation in the biological sense but also statements of general causality between events: *Regulation* events are used generally to connect entities or events stated to other events that they are stated to cause. Thus, PTM events with a stated cause (e.g. a catalyst) can be alternatively represented with a Cause argument on the PTM event or using a separate *Regulation* event (Figure 1). The interpretation of these event structures is identical, and from an annotation perspective there are advantages to both. However, for the purpose of automatic extraction it is important to establish a consistent representation, and thus only one should be used.

In this work, we follow the latter representation, disallowing Cause arguments for annotated PTM events and applying separate Regulation events to capture e.g. catalyst associations. This choice has the benefits of providing an uniform representation for catalysis and inhibition (one involving a Positive regulation and the other a Negative regulation event), reducing the sparseness of specific event structures in the data, and matching the representation chosen in the BioNLP shared task, thus maintaining compatibility with existing event extraction methods. Finally, we note that while we initially expected that glycosylation statements might frequently identify specific attached side chains, necessitating the introduction of an additional argument type to accurately capture all the stated information regarding Glycosylation events, the data contained too few examples for either training material or to justify the modification of the event model. We adopt the constraints applied in the BioNLP shared task regarding the entity types allowed as specific arguments. Thus, the representation we apply here annotated PTM events with specific types, taking as Theme argument a gene/gene product type entity and as Site argument a physical (non-event) entity that does not need to be assigned a specific type.

2.4 Annotation criteria

To create PTM annotation compatible with the event extraction systems introduced for the BioNLP shared task, we created annotation following the GENIA Event corpus annotation criteria (Kim et al., 2008), as adapted for the shared task. The criteria specify that annotation should be applied to statements that involve the occurrence of a change in the state of an entity – even if stated as having occurred in the past, or only hypothetically – but not in cases merely discussing the state or properties of entities, even if these can serve as the basis for inference that a specific change has occurred. We found that many of the spans annotated in PIR as evidence for PTM did not fulfill the criteria for event annotation. The most frequent class consisted of cases where the only evidence for a PTM was in the form of a sequence of residues, for example

> Characterization [...] gave the following sequence, Gly-Cys-Hyp-D-Trp-Glu-Pro-Trp-Cys-NH2 where Hyp = 4-trans-hydroxyproline. (PMID 8910408)

Here, the occurrence of hydroxyproline in the sequence implies that the protein has been hydroxylated, but as the hydroxylation event is only implied by the protein state, no event is annotated.

Candidates drawn from PIR but not fulfilling the criteria were excluded from annotation. While this implies that the general class of event extraction approaches considered here will not recover all statements providing evidence of PTM to biologists (per the PIR criteria), several factors mitigate this limitation of their utility. First, while PTMs implied by sequence only are relatively frequent in PIR, its selection criteria give emphasis to publications initially reporting the existence of a PTM, and further publications discussing the PTM are not expected to state it as sequence only. Thus, it should be possible to extract the corresponding PTMs from later sources. Similarly, one of the promises of event extraction approaches is the potential to extract associations of multiple entities and extract causal chains connecting events with others (e.g. *E catalyzes the hydroxylation of P, leading to ...*), and the data indicates that the sequence-only statements typically provide little information on the biological context of the modification beyond identifying the entity and site. As such non-contextual PTM information is already available in multiple databases, this class of statements may not be of primary interest for event extraction.

2.5 Annotation results

The new PTM annotation covers 157 PubMed abstracts. Following the model of the BioNLP shared task, all mentions of specific gene or gene product names in the abstracts were annotated, applying the annotation criteria of (Ohta et al., 2009). This new named entity annotation covers 1031 gene/gene product mentions, thus averaging more than six mentions per annotated abstract. In total, 422 events of which 405 are of the novel PTM

Event type	Count
Glycosylation	122
Hydroxylation	103
Methylation	90
Acetylation	90
Positive reg.	12
Phosphorylation	3
Protein modification	2
TOTAL	422

Table 6: Statistics of the introduced event annotation.

Arguments	Count
Theme, Site	363
Theme	36
Site	6

Table 7: Statistics for the arguments of the annotated PTM events.

types were annotated, matching the initial annotation target in number and giving a well-balanced distribution of the specific PTM types (Table 6). Reflecting the selection of the source texts, the argument structures of the annotated PTM events (Table 7) show a different distribution from those annotated in the GENIA event corpus (Table 5): whereas less than half of the GENIA event corpus PTM events include a Site argument, almost 90% of the PTM events in the new data include a Site. PTM events identifying both the modified protein and the specific modified site are expected to be of more practical interest. However, we note that the greater number of multi-argument events is expected to make the dataset more challenging as an extraction target.

3 Evaluation

To estimate the capacity of the newly annotated resource to support the extraction of the targeted PTM events and the performance of current event extraction methods at open-domain PTM extraction, we performed a set of experiments using an event extraction method competitive with the state of the art, as established in the BioNLP shared task on event extraction (Kim et al., 2009a; Björne et al., 2009).

3.1 Methods

We adopted the recently introduced event extraction system of Miwa et al. (2010). The system applies a pipeline architecture consisting of three supervised classification-based modules: a trigger detector, an event edge detector, and an event detector. In evaluation on the BioNLP shared task test data, the system extracted phosphorylation events at 75.7% precision and 85.2% recall (80.1% F-score) for Task 1, and 75.7% precision and 83.3% recall (79.3% F-score) for Task 2, showing performance comparable to the best results reported in the literature for this event class (Buyko et al., 2009). We assume three preconditions for the PTM extraction: proteins are given, all PTMs have Sites, and all arguments in a PTM co-occur in sentence scope. The first of these is per the BioNLP shared task setup, the second fixed based the corpus statistics, and the third a property intrinsic to the extraction method, which builds on analysis of sentence structure.[6] In the experiments reported here, only the four novel PTM event types with Sites in the corpus are regarded as a target for the extraction.

The system extracted PTMs as follows: the trigger detector detected the entities (triggers and sites) of the PTMs, the event edge detector detected the edges in the PTMs, and the event detector detected the PTMs. The evaluation setting was the same as the evaluation in (Miwa et al., 2010) except for the threshold. The thresholds in the three modules were tuned with the development data set.

Performance evaluation is performed using the BioNLP shared task primary evaluation criteria, termed the "Approximate Span Matching" criterion. This criterion relaxes the requirements of strict matching in accepting extracted event triggers and entities as correct if their span is inside the region of the corresponding region in the gold standard annotation.

3.2 Data Preparation

The corpus data was split into training and test sets on the document level with a sampling strategy that aimed to preserve a roughly 3:1 ratio of occurrences of each event type between training and test data. The test data was held out during system development and parameter selection and only applied in a single final experiment. The event extraction system was trained using the 112 abstracts of the training set, further using 24 of the abstracts

[6] We note that in the BioNLP shared task data, all arguments were contained within single sentences for 95% of events.

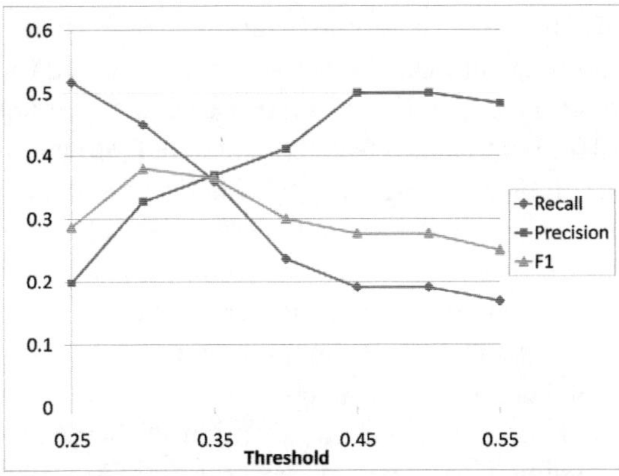

Figure 2: Performance of PTM extraction on the development data set.

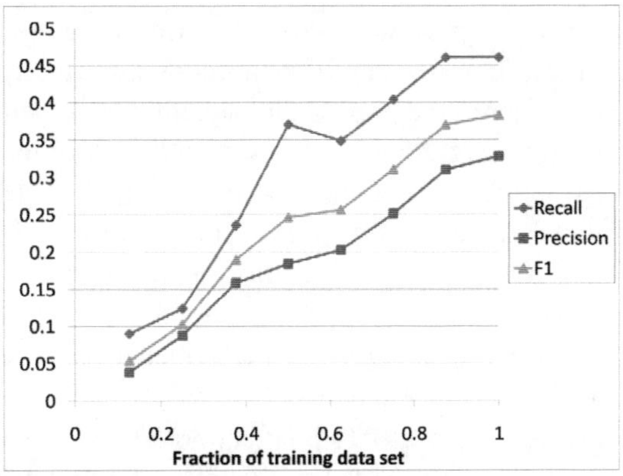

Figure 3: Learning curve of PTM extraction on the development data set.

Event type	Prec	Rec	F
Acetylation	69.6%	36.7%	48.1%
Methylation	50.0%	34.2%	40.6%
Glycosylation	36.7%	42.5%	39.4%
Hydroxylation	57.1%	29.3%	38.7%
Overall	52.1%	35.7%	42.4%

Table 8: Event extraction results on the test set.

as a development test set.

3.3 Results

We first performed parameter selection, setting the machine learning method parameter by estimating performance on the development data set. Figure 2 shows the performance of PTM extraction on the development data set with different values of parameter. The threshold value corresponding to the best performance (0.3) was then applied for an experiment on the held-out test set.

Performance on the test set was evaluated as 52% precision and 36% recall (42% F-score), matching estimates on the development data. A breakdown by event type (Table 8) shows that *Acetylation* is most reliably extracted with extraction for the other three PTM types showing similar F-scores despite some variance in the precision/recall balance. We note that while these results fall notably below the best result reported for Phosphorylation events in the BioNLP shared task, they are comparable to the best results reported in the task for Regulation and Binding events (Kim et al., 2009a), suggesting that the dataset allows the extraction of the novel PTM events with Theme and Site arguments at levels comparable to multi-argument shared task events.

Further, a learning curve (Figure 3) plotted on the development data suggests roughly linearly increasing performance over most of the curve. While the increase appears to be leveling off to an extent when using all of the available data, the learning curve indicates that performance can be further improved by increasing the size of the annotated dataset.

4 Discussion

Post-translational modifications have been a focus of interest in the biomedical text mining community, and a number of resources and systems targeting PTM have been proposed. The GENIES and GeneWays systems (Friedman et al., 2001; Rzhetsky et al., 2004) targeted PTM events such as phosphorylation and dephosphorylation under the more general *createbond* and *breakbond* types. Hu et al. (2005) introduce the RLIMS-P rule-based system for mining the substrates and sites for phosphorylation, which is extended with the capacity to extract intra-clausal statements by Narayanaswamy et al. (2005). Saric et al. (2006) present an extension of their rule-based STRING-IE system for extracting regulatory networks to capture phosphorylation and dephosphorylation events. Lee et al. (2008) present E3Miner, a tool for automatically extracting information related to ubiquitination, and Kim et al. (2009b) present a preliminary study adapting the E3Miner approach to the mining of acetylation events.

It should be noted that while studies targeting single specific PTM types report better results than found in the initial evaluation presented here (in many cases dramatically so), different extraction targets and evaluation criteria complicate direct comparison. Perhaps more importantly, our aim here is to extend the capabilities of general event extraction systems targeting multiple types of structured events. Pursuing this broader goal necessarily involves some compromise in the ability to focus on the extraction of individual event types, and it is expected that highly focused systems will provide better performance than retrained general systems.

The approach to PTM extraction adopted here relies extensively on the availability of annotated resources, the creation of which requires considerable effort and expertise in understanding the target domain as well as the annotation methodology and tools. The annotation created in this study, performed largely on the basis of partial existing annotations drawn from PIR data, involved an estimated three weeks of full-time effort from an experienced annotator. As experiments further indicated that a larger corpus may be necessary for reliable annotation, we can estimate that extending the approach to sufficient coverage of each of hundreds of PTM types without a partial initial annotation would easily require several person-years of annotation efforts. We thus see a clear need for the development of unsupervised or semisupervised methods for PTM extraction to extend the coverage of event extraction systems to the full scale of different PTM types. Nevertheless, even if reliable methods for PTM extraction that entirely avoid the need for annotated training data become available, a manually curated reference standard will still be necessary for reliable estimation of their performance. To efficiently support the development of event extraction systems capable of capturing the full variety of PTM events, it may be beneficial to reverse the approach taken here: instead of annotating hundreds of examples of a small number of PTM types, annotate a small number of each of hundreds of PTM types, thus providing both seed data for semisupervised approaches as well as reference data for the evaluation of broad-coverage PTM event extraction systems.

5 Conclusions and Future Work

We have presented an event extraction approach to automatic PTM recognition, building on the model introduced in the BioNLP shared task on event extraction. By annotating a targeted corpus for four prominent PTM types not considered in the BioNLP shared task data, we have created a resource that can be straightforwardly used to extend the capability of event extraction systems for PTM extraction. We estimated that while systems trained on the original shared task dataset could not recognize more than 50% of PTM mentions due to their types, the introduced annotation increases this theoretical upper bound to nearly 90%. An initial experiment on the newly introduced dataset using a state-of-the-art method indicated that straightforward adoption of the dataset as training data to extend coverage of PTM events without specific adaptations of the method is feasible, although the measured performance indicates remaining challenges for reliable extraction. Further, while the experiments were performed on a dataset selected to avoid bias toward e.g. a particular subdomain or specific forms of event expressions, it remains an open question how extraction performance generalizes to biomedical literature beyond the selected sample. As experiments indicated clear remaining potential for the improvement of extraction performance from more training data, the extension of the annotated dataset is a natural direction for future work. We considered also the possiblity of extending annotation to cover small numbers of each of a large variety of PTM types, which would place focus on the challenges of event extraction with little or no training data for specific event types.

The annotated corpus covering over 1000 gene and gene product entities and over 400 events is freely available in the widely adopted BioNLP shared task format at the GENIA project homepage.[7]

Acknowledgments

We would like to thank Goran Topic for automating Medie queries to identify target abstracts. This work was partially supported by Grant-in-Aid for Specially Promoted Research (MEXT, Japan) and Japan-Slovenia Research Cooperative Program (JSPS, Japan and MHEST, Slovenia).

References

Sophia Ananiadou, Sampo Pyysalo, Junichi Tsujii, and Douglas B. Kell. 2010. Event extraction for systems biology by text mining the literature. *Trends in Biotechnology*. (to appear).

[7] http://www-tsujii.is.s.u-tokyo.ac.jp/GENIA

Jari Björne, Juho Heimonen, Filip Ginter, Antti Airola, Tapio Pahikkala, and Tapio Salakoski. 2009. Extracting complex biological events with rich graph-based feature sets. In *Proceedings of the BioNLP 2009 Workshop Companion Volume for Shared Task*, pages 10–18, Boulder, Colorado, June. Association for Computational Linguistics.

Ekaterina Buyko, Erik Faessler, Joachim Wermter, and Udo Hahn. 2009. Event extraction from trimmed dependency graphs. In *Proceedings of the BioNLP 2009 Workshop Companion Volume for Shared Task*, pages 19–27, Boulder, Colorado, June. Association for Computational Linguistics.

Carol Friedman, Pauline Kra, Hong Yu, Michael Krauthammer, and Andrey Rzhetsky. 2001. GENIES: A natural-language processing system for the extraction of molecular pathways from journal articles. *Bioinformatics*, 17(Suppl. 1):S74–S82.

Z. Z. Hu, M. Narayanaswamy, K. E. Ravikumar, K. Vijay-Shanker, and C. H. Wu. 2005. Literature mining and database annotation of protein phosphorylation using a rule-based system. *Bioinformatics*, 21(11):2759–2765.

Rudolf Jaenisch and Adrian Bird. 2003. Epigenetic regulation of gene expression: how the genome integrates intrinsic and environmental signals. *Nature Genetics*, 33:245–254.

Jin-Dong Kim, Tomoko Ohta, and Jun'ichi Tsujii. 2008. Corpus annotation for mining biomedical events from lterature. *BMC Bioinformatics*, 9(1):10.

Jin-Dong Kim, Tomoko Ohta, Sampo Pyysalo, Yoshinobu Kano, and Jun'ichi Tsujii. 2009a. Overview of bionlp'09 shared task on event extraction. In *Proceedings of the BioNLP 2009 Workshop Companion Volume for Shared Task*, pages 1–9, Boulder, Colorado, June. Association for Computational Linguistics.

Youngrae Kim, Hodong Lee, and Gwan-Su Yi. 2009b. Literature mining for protein acetylation. In *Proceedings of LBM'09*.

Hodong Lee, Gwan-Su Yi, and Jong C. Park. 2008. E3Miner: a text mining tool for ubiquitin-protein ligases. *Nucl. Acids Res.*, 36(suppl.2):W416–422.

Matthias Mann and Ole N. Jensen. 2003. Proteomic analysis of post-translational modifications. *Nature Biotechnology*, 21:255–261.

Makoto Miwa, Rune Sætre, Jin Dong Kim, and Jun'ichi Tsujii. 2010. Event extraction with complex event classification using rich features. *Journal of Bioinformatics and Computational Biology (JBCB)*, 8(1):131–146, February.

Yusuke Miyao, Tomoko Ohta, Katsuya Masuda, Yoshimasa Tsuruoka, Kazuhiro Yoshida, Takashi Ninomiya, and Jun'ichi Tsujii. 2006. Semantic Retrieval for the Accurate Identification of Relational Concepts in Massive Textbases. In *Proceedings of COLING-ACL 2006*, pages 1017–1024.

M. Narayanaswamy, K. E. Ravikumar, and K. Vijay-Shanker. 2005. Beyond the clause: extraction of phosphorylation information from medline abstracts. *Bioinformatics*, 21(suppl.1):i319–327.

Tomoko Ohta, Yusuke Miyao, Takashi Ninomiya, Yoshimasa Tsuruoka, Akane Yakushiji, Katsuya Masuda, Jumpei Takeuchi, Kazuhiro Yoshida, Tadayoshi Hara, Jin-Dong Kim, Yuka Tateisi, and Jun'ichi Tsujii. 2006. An Intelligent Search Engine and GUI-based Efficient MEDLINE Search Tool Based on Deep Syntactic Parsing. In *Proceedings of the COLING/ACL 2006 Interactive Presentation Sessions*, pages 17–20.

Tomoko Ohta, Jin-Dong Kim, Sampo Pyysalo, and Jun'ichi Tsujii. 2009. Incorporating GENETAG-style annotation to GENIA corpus. In *Proceedings of Natural Language Processing in Biomedicine (BioNLP) NAACL 2009 Workshop*, pages 106–107, Boulder, Colorado. Association for Computational Linguistics.

Sampo Pyysalo, Filip Ginter, Juho Heimonen, Jari Björne, Jorma Boberg, Jouni Järvinen, and Tapio Salakoski. 2007. BioInfer: A corpus for information extraction in the biomedical domain. *BMC Bioinformatics*, 8(50).

Andrey Rzhetsky, Ivan Iossifov, Tomohiro Koike, Michael Krauthammer, Pauline Kra, Mitzi Morris, Hong Yu, Pablo Ariel Duboué, Wubin Weng, W. John Wilbur, Vasileios Hatzivassiloglou, and Carol Friedman. 2004. GeneWays: A system for extracting, analyzing, visualizing, and integrating molecular pathway data. *Journal of Biomedical Informatics*, 37(1):43–53.

Jasmin Saric, Lars Juhl Jensen, Rossitza Ouzounova, Isabel Rojas, and Peer Bork. 2006. Extraction of regulatory gene/protein networks from Medline. *Bioinformatics*, 22(6):645–650.

The Gene Ontology Consortium. 2000. Gene ontology: tool for the unification of biology. *Nature Genetics*, 25:25–29.

Eric S Witze, William M Old, Katheryn A Resing, and Natalie G Ahn. 2007. Mapping protein post-translational modifications with mass spectrometry. *Nature Methods*, 4:798–806.

Cathy H. Wu, Lai-Su L. Yeh, Hongzhan Huang, Leslie Arminski, Jorge Castro-Alvear, Yongxing Chen, Zhangzhi Hu, Panagiotis Kourtesis, Robert S. Ledley, Baris E. Suzek, C.R. Vinayaka, Jian Zhang, and Winona C. Barker. 2003. The Protein Information Resource. *Nucl. Acids Res.*, 31(1):345–347.

Scaling up Biomedical Event Extraction to the Entire PubMed

Jari Björne[*,1,2] **Filip Ginter**,[*,1] **Sampo Pyysalo**,[*,3] **Jun'ichi Tsujii**,[3,4] **Tapio Salakoski**[1,2]
[1]Department of Information Technology, University of Turku, Turku, Finland
[2]Turku Centre for Computer Science (TUCS), Turku, Finland
[3]Department of Computer Science, University of Tokyo, Tokyo, Japan
[4]National Centre for Text Mining, University of Manchester, Manchester, UK
jari.bjorne@utu.fi, ginter@cs.utu.fi, smp@is.s.u-tokyo.ac.jp
tsujii@is.s.u-tokyo.ac.jp, tapio.salakoski@it.utu.fi

Abstract

We present the first full-scale event extraction experiment covering the titles and abstracts of all PubMed citations. Extraction is performed using a pipeline composed of state-of-the-art methods: the BANNER named entity recognizer, the McClosky-Charniak domain-adapted parser, and the Turku Event Extraction System. We analyze the statistical properties of the resulting dataset and present evaluations of the core event extraction as well as negation and speculation detection components of the system. Further, we study in detail the set of extracted events relevant to the apoptosis pathway to gain insight into the biological relevance of the result. The dataset, consisting of 19.2 million occurrences of 4.5 million unique events, is freely available for use in research at `http://bionlp.utu.fi/`.

1 Introduction

There has recently been substantial interest in *event models* in biomedical information extraction (IE). The expressive event representation captures extracted knowledge as structured, recursively nested, typed associations of arbitrarily many participants in specific roles. The BioNLP'09 Shared Task on Event Extraction (Kim et al., 2009), the first large scale evaluation of biomedical event extraction systems, drew the participation of 24 groups and established a standard event representation scheme and datasets. The training and test data of the Shared Task comprised 13,623 manually annotated events in 1,210 PubMed citation abstracts, and on this data the top performing system of Björne et al. (2009; 2010b) achieved an overall F-score of 51.95% (Kim et al., 2009).

The issue of the scalability and generalization ability of the introduced event extraction systems beyond the domain of the GENIA corpus on which the Shared Task was based has remained largely an open question. In a prior study, we have established on a 1% random sample of PubMed titles and abstracts that the event extraction system of Björne et al. is able to scale up to PubMed-wide extraction without prohibitive computational time requirements, however, the actual extraction from the entire PubMed was left as a future work (Björne et al., 2010a). Thus, the top-ranking event extraction systems in the Shared Task have, in fact, not been used so far for actual mass-scale event extraction beyond the carefully controlled setting of the Shared Task itself. Further, since an automated named entity recognition step was not part of the Shared Task, the interaction of the event extraction systems with gene/protein name recognizers remains largely unexplored as well.

In this study, we address some of these questions by performing a mass-scale event extraction experiment using the best performing system[1] of the Shared Task (Björne et al., 2009; Björne et al., 2010b), and applying it to the entire set of titles and abstracts of the nearly 18 million citations in the 2009 distribution of PubMed. The extraction result, containing 19.2 million event occurrences, is the largest dataset of its type by several orders of magnitude and arguably represents the state-of-the-art in automatic event extraction with respect to both accuracy and size.

To support emerging community efforts in tasks that build on event extraction output, such as event network refinement, hypothesis generation, pathway extraction, and others, we make the entire resulting dataset freely available for research purposes. This allows researchers interested in questions involving *text mining*, rather than initial in-

*Equal contribution by first three authors.

[1]Available at `http://bionlp.utu.fi/`

Proceedings of the 2010 Workshop on Biomedical Natural Language Processing, ACL 2010, pages 28–36,
Uppsala, Sweden, 15 July 2010.

Event type	Example
Gene expression	5-LOX is *expressed* in leukocytes
Transcription	promoter associated with IL-4 gene *transcription*
Localization	phosphorylation and nuclear *translocation* of STAT6
Protein catabolism	I kappa B-alpha *proteolysis* by phosphorylation.
Phosphorylation	BCL-2 was *phosphorylated* at the G(2)/M phase
Binding	Bcl-w *forms complexes* with Bax and Bak
Regulation	c-Met expression is *regulated* by Mitf
Positive regulation	IL-12 *induced* STAT4 binding
Negative regulation	DN-Rac *suppressed* NFAT activation

Table 1: Targeted event types with brief example statements expressing an event of each type. In the examples, the word or words marked as triggering the presence of the event are shown in italics and event participants underlined. The event types are grouped by event participants, with the first five types taking one *theme*, binding events taking multiple *themes* and the regulation types *theme* and *cause* participants. Adapted from (Björne et al., 2009).

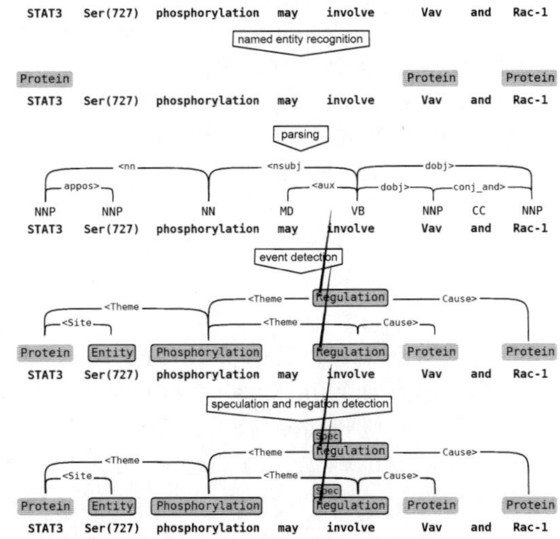

Figure 1: Event extraction. A multi-stage system produces an *event graph* for each sentence. Named entities are detected (A) using BANNER. Independently of named entity detection, sentences are parsed (B) to produce a dependency parse. Event detection (C) uses the named entities and the parse in predicting the *trigger nodes* and *argument edges* that form the events. Finally, polarity and certainty (D) are predicted for the generated events. Adapted from (Björne et al., 2009).

formation extraction, to make use of the many favorable statistical properties of the massive dataset without having to execute the laborious and time-consuming event extraction pipeline.

In the following, we describe the Shared Task event representation applied throughout this study, the event extraction pipeline itself, and a first set of analyzes of multiple aspects of the resulting dataset.

2 Event extraction

The event extraction pipeline follows the model of the BioNLP'09 Shared Task in its representation of extracted information. The primary extraction targets are gene or gene product-related entities and nine fundamental biomolecular event types involving these entities (see Table 1 for illustration).

Several aspects of the event representation, as defined in the context of the Shared Task, differentiate the event extraction task from the body of domain IE studies targeting e.g. protein–protein interactions and gene–disease relations, including previous domain shared tasks (Nédellec, 2005; Krallinger et al., 2008). Events can have an arbitrary number of participants with specified roles (e.g. *theme* or *cause*), making it possible to capture n-ary associations and statements where some participants occur in varying roles or are only occasionally mentioned. A further important property is that event participants can be other events, resulting in expressive, recursively nested structures. Finally, events are given GENIA Event ontology types drawn from the community-standard Gene Ontology (The Gene Ontology Consortium, 2000), giving each event well-defined semantics.

2.1 Event Extraction Pipeline

The event extraction pipeline applied in this work consists of three main processing steps: named entity recognition, syntactic parsing, and event extraction. The process is illustrated in Figure 1.

For named entity recognition, we use the BANNER system of Leaman and Gonzales (2008), which in its current release achieves results close to the best published on the standard GENETAG dataset and was reported to have the best performance in a recent study comparing publicly available taggers (Kabiljo et al., 2009). Titles and abstracts of all 17.8M citations in the 2009 distribution of PubMed are processed through the BANNER system.

Titles and abstracts of PubMed citations in which at least one named entity was identified, and which therefore contain a possible target for event extraction, are subsequently split into sentences using a maximum-entropy based sentence splitter trained on the GENIA corpus (Kazama and Tsujii, 2003) with limited rule-based post-processing for some common errors.

All sentences containing at least one named entity are then parsed with the domain-adapted McClosky-Charniak parser (McClosky and Charniak, 2008; McClosky, 2009), which has achieved the currently best published performance on the GENIA Treebank (Tateisi et al., 2005). The constituency parse trees are then transformed to the *collapsed-ccprocessed* variant of the Stanford Dependency scheme using the conversion tool[2] introduced by de Marneffe et al. (2006).

Finally, events are extracted using the Turku Event Extraction System of Björne et al. which achieved the best performance in the BioNLP'09 Shared Task and remains fully competitive with even the most recent advances (Miwa et al., 2010). We use a recent publicly available revision of the event extraction system that performs also extraction of Shared Task subtask 2 and 3 information, providing additional event arguments relevant to event sites and localization (*site*, *atLoc*, and *toLoc* role types in the Shared Task) as well as information on event polarity and certainty (Björne et al., 2010b).

2.2 Extraction result and computational requirements

Named entity recognition using the BANNER system required in total roughly 1,800 CPU-hours and resulted in 36,454,930 named entities identified in 5,394,350 distinct PubMed citations.

Parsing all 20,037,896 sentences with at least one named entity using the McClosky-Charniak parser and transforming the resulting constituency trees into dependency analyzes using the Stanford conversion tool required about 5,000 CPU-hours, thus averaging 0.9 sec per sentence. Even though various stability and scalability related problems were met during the parsing process, we were able to successfully parse 20,020,266 (99.91%) of all sentences.

Finally, the event extraction step required approximately 1,500 CPU-hours and resulted in 19,180,827 event instances. In total, the entire corpus of PubMed titles and abstracts was thus processed in roughly 8,300 CPU-hours, or, 346 CPU-days, the most time-consuming step by far being the syntactic parsing.

We note that, even though the components used in the pipeline are largely well-documented and mature, a number of technical issues directly related to, or at least magnified by, the untypically large dataset were met at every point of the pipeline. Executing the pipeline was thus far from a trivial undertaking. Due to the computational requirements of the pipeline, cluster computing systems were employed at every stage of the process.

2.3 Evaluation

We have previously evaluated the Turku Event Extraction System on a random 1% sample of PubMed citations, estimating a precision of 64% for event types and arguments pertaining to subtask 1 of the Shared Task (Björne et al., 2010a), which compares favorably to the 58% precision the system achieves on the Shared Task dataset itself (Björne et al., 2009).

To determine precision on subtasks 2 and 3 on PubMed citations, we manually evaluate 100 events with *site* and *location* arguments (subtask 2) and 100 each of events predicted to be *speculated* or *negated* (subtask 3).

Subtask 2 *site* and *location* arguments are mostly external to the events they pertain to and therefore were evaluated independently of their parent event. Their precision is 53% (53/100), comparable to the 58% precision established on the BioNLP'09 Shared Task development set, using the same parent-independent criterion.

To estimate the precision of the negation detection (subtask 3), we randomly select 100 events predicted to be negated. Of these, 9 were incorrect as events to such an extent that the correctness of the predicted negation could not be judged and, among the remaining 91 events, the negation was correctly predicted in 82% of the cases. Similarly, to estimate the precision of speculation detection, we randomly select 100 events predicted to be speculated, of which 20 could not be judged for correctness of speculation. Among the remaining 80, 88% were correctly predicted as speculative events. The negations were mostly signalled by explicit statements such as *is not regulated*, and speculation by statements, such as *was studied*, that defined the events as experimental questions.

[2] http://www-nlp.stanford.edu/downloads/lex-parser.shtml

For comparison, on the BioNLP'09 Shared Task development set, for correctly predicted events, precision for negation examples was 83% (with recall of 53%) and for speculation examples 77% (with recall of 51%).

In the rest of this paper, we turn our attention to the extraction result.

3 Term-NE mapping

As the event types are drawn from the Gene Ontology and the original data on which the system is trained has been annotated with reference to the GO definitions, the events targeted by the extraction system have well-defined biological interpretations. The meaning of complete event structures depends also on the participating entities, which are in the primary event extraction task constrained to be of gene/gene product (GGP) types, as annotated in the GENIA GGP corpus (Ohta et al., 2009a). The simple and uniform nature of these entities makes the interpretation of complete events straightforward.

However, the semantics of the entities automatically tagged in this work are somewhat more openly defined. The BANNER system was trained on the GENETAG corpus, annotated for "gene/protein entities" without differentiating between different entity types and marking entities under a broad definition that not only includes genes and gene products but also related entities such as gene promoters and protein complexes, only requiring that the tagged entities be specific (Tanabe et al., 2005). The annotation criteria of the entities used to train the BANNER system as well as the event extraction system also differ in the extent of the marked spans, with GENIA GGP marking the minimal name and GENETAG allowing also the inclusion of head nouns when a name occurs in modifier position. Thus, for example, the latter may annotate the spans *p53 gene*, *p53 protein*, *p53 promoter* and *p53 mutations* in contexts where the former would in each case mark only the substring *p53*.

One promising future direction for the present effort is to refine the automatically extracted data into an event network connected to specific entries in gene/protein databases such as Entrez Gene and UniProt. To achieve this goal, the resolution of the tagged entities can be seen to involve two related but separate challenges. First, identifying the specific database entries that are referred to by the genes/proteins named in the tagged entities, and second, mapping from the events involving automatically extracted terms to ones involving the associated genes/proteins. The first challenge, gene/protein name normalization, is a well-studied task in biomedical NLP for which a number of systems with promising performance have been proposed (Morgan and Hirschman, 2007). The second we believe to be novel. In the following, we propose a method for resolving this task.

Relation	Examples
Equivalent	GGP gene, wild-type GGP
Class-Subclass	human GGP, HIV-1 GGP
Object-Variant	
GGP-Isoform	GGP isoform
GGP-Mutant	dominant-negative GGP
GGP-Recombinant	GGP expression plasmid
GGP-Precursor	GGP precursor, pro-GGP
Component-Object	
GGP-Amino acid	GGP-Ile 729
GGP-AA motif	GGP NH2-terminal
GGP-Reg. element	GGP proximal promoter
GGP-Flanking region	GGP 5' upstream sequence
Object-Component	
GGP-Protein Complex	GGP homodimers
Place-Area	
GGP-Locus	GGP loci
Member-Collection	
GGP-Group	GGP family members

Table 2: Gene/gene product NE-term relation types with examples. Top-level relations in the relation type hierarchy shown in bold, specific NE names in examples replaced with *GGP*. Intermediate levels in the hierarchy and a number of minor relations omitted. Relation types judged to allow remapping (see text) underlined.

We base the decision on how to map events referencing broadly defined terms to ones referencing associated gene/protein names in part on a recently introduced dataset of "static relations" (Pyysalo et al., 2009) between named entities and terms (Ohta et al., 2009b). This dataset was created based on approximately 10,000 cases where GGP NEs, as annotated in the GENIA GGP corpus (Ohta et al., 2009a), were embedded in terms, as annotated in the GENIA term corpus (Ohta et al., 2002). For each such case, the relation between the NE and the term was annotated using a set of introduced relation types whose granularity was defined with reference to MeSH terms (see Table 2, Ohta et al., 2009b). From this data, we extracted prefix and suffix strings that, when affixed to a GGP name, produced a term with a predictable relation (within the dataset) to the GGP. Thus, for example, the

term	GGP
p53 protein	p53
p53 gene	p53
human serum albumin	serum albumin
wild-type p53	p53
c-fos mRNA	c-fos
endothelial NO synthase	NO synthase
MHC cl. II molecules	MHC cl. II
human insulin	insulin
HIV-1 rev.transcriptase	rev.transcriptase
hepatic lipase	lipase
p24 antigen	p24
tr. factor NF-kappaB	NF-kappaB
MHC molecules	MHC
PKC isoforms	PKC
HLA alleles	HLA
RET proto-oncogene	RET
ras oncogene	ras
SV40 DNA	SV40
EGFR tyrosine kinase	EGFR

Table 3: Examples of frequently applied mappings. Most frequent term for each mapping is shown. Some mention strings are abbreviated for space.

	Mentions	Types
Total	36454930	4747770
Mapped	2212357 (6.07%)	547920 (11.54%)
Prefix	430737 (1.18%)	129536 (2.73%)
Suffix	1838646 (5.04%)	445531 (9.38%)

Table 4: Statistics for applied term-GGP mappings. Tagged mentions and types (unique mentions) shown separately. Overall total given for reference, for mappings overall for any mapping shown and further broken down into prefix-string and suffix-string based.

prefix string "wild-type" was associated with the *Equivalent* relation type and the suffix string "activation sequence" with the *GGP-Regulatory element* type. After filtering out candidates shorter than 3 characters as unreliable (based on preliminary experiments), this procedure produced a set of 68 prefix and 291 suffix strings.

To make use of the data for predicting relations between GGP names and the terms formed by affixing a prefix or suffix string, it is necessary to first identify name-term pairs. Candidates can be generated simply by determining the prefix/suffix strings occurring in each automatically tagged entity and assuming that what remains after removing the prefixes and suffixes is a GGP name. However, this naive strategy often fails: while removing "protein" from "p53 protein" correctly identifies "p53" as the equivalent GGP name, for "capsid protein" the result, "capsid" refers not to a GGP but to the shell of a virus – "protein" is properly part of the protein name. To resolve this issue, we drew on the statistics of the automatically tagged entities, assuming that if a prefix/suffix string is not a fixed part of a name, the name will appear tagged also without that string. As the tagging covers the entire PubMed, this is likely to hold for all but the very rarest GGP names. To compensate for spurious hits introduced by tagging errors, we specifically required that to accept a candidate prefix/suffix string-name pair, the candidate name should occur more frequently without the prefix/suffix than with it. As the dataset is very large, this simple heuristic often gives the right decision with secure margins: for example, "p53" was tagged 117,835 times but "p53 protein" only 11,677, while "capsid" was (erroneously) tagged 7 times and "capsid protein" tagged 1939 times.

A final element of the method is the definition of a mapping to events referencing GGP NEs from the given events referencing terms, the NEs contained in the terms, and the NE-term relations. In this work, we apply independently for each term a simple mapping based only on the relation types, deciding for each type whether replacing reference to a term with reference to a GGP holding the given relation to the term preserves event semantics (to an acceptable approximation) or not. For the Equivalent relation this holds by definition. We additionally judged all Class-Subclass and Component-Object relations to allow remapping (accepting e.g. P_1 *binds part of* $P_2 \rightarrow P_1$ *binds* P_2) as well as selected Object-Variant relations (see Table 2). For cases judged not to allow remapping, we simply left the event unmodified.

Examples of frequently applied term-GGP mappings are shown in Table 3, and Table 4 shows the statistics of the applied mappings. We find that suffix-based mappings apply much more frequently than prefix-based, perhaps reflecting also the properties of the source dataset. Overall, the number of unique tagged types is reduced by over 10% by this procedure. It should be noted that the applicability of the method could likely be considerably extended by further annotation of NE-term relations in the dataset of Ohta et al. (2009b): the current data is all drawn from the GENIA corpus, drawn from the subdomain of transcription factors in human blood cells, and its coverage of PubMed is thus far from exhaustive.

4 Event recurrence

Given a dataset of events extracted from the entire PubMed, we can study whether, and to what extent, events are re-stated in multiple PubMed citations. This analysis may shed some light — naturally within the constraints of an automatically extracted dataset rather than gold-standard annotation — on the often (informally) discussed hypothesis that a high-precision, low recall system might be a preferred choice for large-scale extraction as the lower recall would be compensated by the redundancy of event statements in PubMed.

In order to establish event recurrence statistics, that is, the number of times a given event is repeated in the corpus, we perform a limited normalization of tagged entities consisting of the Term-NE mapping presented in Section 3 followed by lowercasing and removal of non-alphanumeric characters. Two named entities are then considered equal if their normalized string representations are equal. For instance, the two names *IL-2 gene* and *IL2* would share the same normalized form *il2* and would thus be considered equal.

For the purpose of recurrence statistics, two events are considered equal if their types are equal, and all their Theme and Cause arguments, which can be other events, are recursively equal as well. A canonical order of arguments is used in the comparison, thus e.g. the following events are considered equal:

regulation(Cause:A, Theme:binding(Theme:B, Theme:C))
regulation(Theme:binding(Theme:C, Theme:B), Cause:A)

In total, the system extracted 19,180,827 instances of 4,501,883 unique events. On average, an event is thus stated 4.2 times. The distribution is, however, far from uniform and exhibits the "long tail" typical of natural language phenomena, with 3,484,550 (77%) of events being singleton occurrences. On the other hand, the most frequent event, *localization(Theme:insulin)*, occurs as many as 59,821 times. The histogram of the number of unique events with respect to their occurrence count is shown in Figure 2.

The total event count consists mostly of simple one-argument events. The arguably more interesting category of events that involve at least two different named entities constitutes 2,064,278 instances (11% of the 19.2M total) of 1,565,881 unique events (35% of the 4.5M total). Among these complex events, recurrence is thus considerably lower, an event being stated on average 1.3 times. The most frequent complex event, with 699 occurrences, is *positive-regulation(Cause:GnRG,Theme:localization(Theme:LH))*, reflecting the well-known fact that *GnRG* causes the release of *LH*, a hormone important in human reproduction.

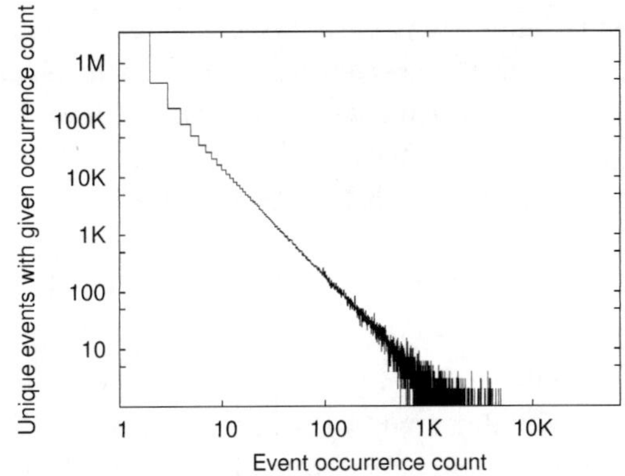

Figure 2: Number of unique events (y-axis) with a given occurrence count (x-axis).

	R	P	N	L	B	E	T	C	H
R	**561**	173	128	42	63	83	30	16	17
P	173	**1227**	192	58	99	143	39	20	23
N	128	192	**668**	46	73	98	31	17	18
L	42	58	46	**147**	57	75	25	15	15
B	63	99	73	57	**1023**	134	35	20	21
E	83	143	98	75	134	**705**	49	22	24
T	30	39	31	25	35	49	**79**	11	11
C	16	20	17	15	20	22	11	**39**	7
H	17	23	18	15	21	24	11	7	**49**

Table 5: Event type confusion matrix. Each element contains the number of unique events, in thousands, that are equal except for their type. The matrix is symmetric and its diagonal sums to 4,5M, the total number of extracted unique events. The event types are (R)egulation, (P)ositive regulation, (N)egative regulation, (L)ocalization, (B)inding, gene (E)xpression, (T)ranscription, protein (C)atabolism, and p(H)osphorylation.

To gain an additional broad overview of the characteristics of the extracted events, we compute an *event type confusion matrix*, shown in Table 5. In this matrix, we record for each pair of event types T_1 and T_2 the number of unique events of type T_1 for which an event of type T_2 can be found such that, apart for the type difference, the events are otherwise equal. While e.g. a positive regulation-negative regulation pair is at least unusual, in general these event pairs do not suggest extraction errors: for instance the existence of the event *expression(Theme:A)* does not in any way prevent the existence of the event *localization(Theme:A)*, and *regulation* subsumes *positive-regulation*. Nevertheless, Table 5 shows a clear preference for a single type for the events.

5 Case Study: The apoptosis pathway

In this section, we will complement the preceding broad statistical overview of the extracted events with a detailed study of a specific pathway, the *apoptosis pathway*, determining how well the extracted events cover its interactions (Figure 3).

To create an event network, the events must be linked through their protein arguments. In addition to the limited named entity normalization introduced in Section 4, we make use of a list of synonyms for each protein name in the apoptosis pathway, obtained manually from protein databases, such as UniProt. Events whose protein arguments correspond to any of these known synonyms are then used for reconstructing the pathway.

The apoptosis pathway consists of several overlapping signaling routes and can be defined on different levels of detail. To have a single, accurate and reasonably high-level definition, we based our pathway on a concisely presentable subset of the KEGG human apoptosis pathway (entry hsa04210) (Kanehisa and Goto, 2000). As seen in Figure 3, the extracted dataset contains events between most interaction partners in the pathway.

The constructed pathway also shows that the extracted events are not necessarily interactions in the physical sense. Many "higher level" events are extracted as well. For example, the extracellular signaling molecule *TNF*α can trigger pathways leading to the activation of *Nf-*κ*B*. Although the two proteins are not likely to interact directly, it can be said that *TNF*α upregulates *NF-*κ*B*, an event actually extracted by the system. Such statements of indirect interaction co-exist with statements of actual, physical interactions in the event data.

6 Conclusions

In this paper, we have presented the result of processing the entire, unabridged set of PubMed titles and abstracts with a state-of-the-art event extraction pipeline as a new resource for text mining in the biomedical domain. The extraction result arguably represents the best event extraction output achievable with currently available tools.

The primary contribution of this work is the set of over 19M extracted event instances of 4.5M unique events. Of these, 2.1M instances of 1.6M unique events involve at least two different named entities. These form an event network several orders of magnitude larger than those previously available. The data is intended to support research in biological hypothesis generation, pathway extraction, and similar higher-level text mining tasks. With the network readily available in an easy-to-process format under an open license, researchers can focus on the core tasks of text mining without the need to perform the tedious and computationally very intensive task of event extraction with a complex IE pipeline.

In addition to the extracted events, we make readily available the output of the BANNER system on the entire set of PubMed titles and abstracts as well as the parser output of the McClosky-Charniak domain-adapted parser (McClosky and Charniak, 2008; McClosky, 2009) further transformed to the Stanford Dependency representation using the tools of de Marneffe et al. (2006) for nearly all (99.91%) sentences with at least one named entity identified. We expect this data to be of use for the development and application of systems for event extraction and other BioNLP tasks, many of which currently make extensive use of dependency syntactic analysis. The generation of this data having been far from a trivial technical undertaking, its availability as-is can be expected to save substantial duplication of efforts in further research.

A manual analysis of extracted events relevant to the apoptosis pathway demonstrates that the event data can be used to construct detailed biological interaction networks with reasonable accuracy. However, accurate entity normalization, in particular taking into account synonymous names, seems to be a necessary prerequisite and remains among the most important future work directions. In the current study, we take first steps in this direction in the form of a term-NE mapping method in event context. The next step will be the application of a state-of-the-art named entity normalization system to obtain biological database identities for a number of the named entities in the extracted event network, opening possibilities for combining the data in the network with other biological information. A further practical problem to address will be that of visualizing the network and

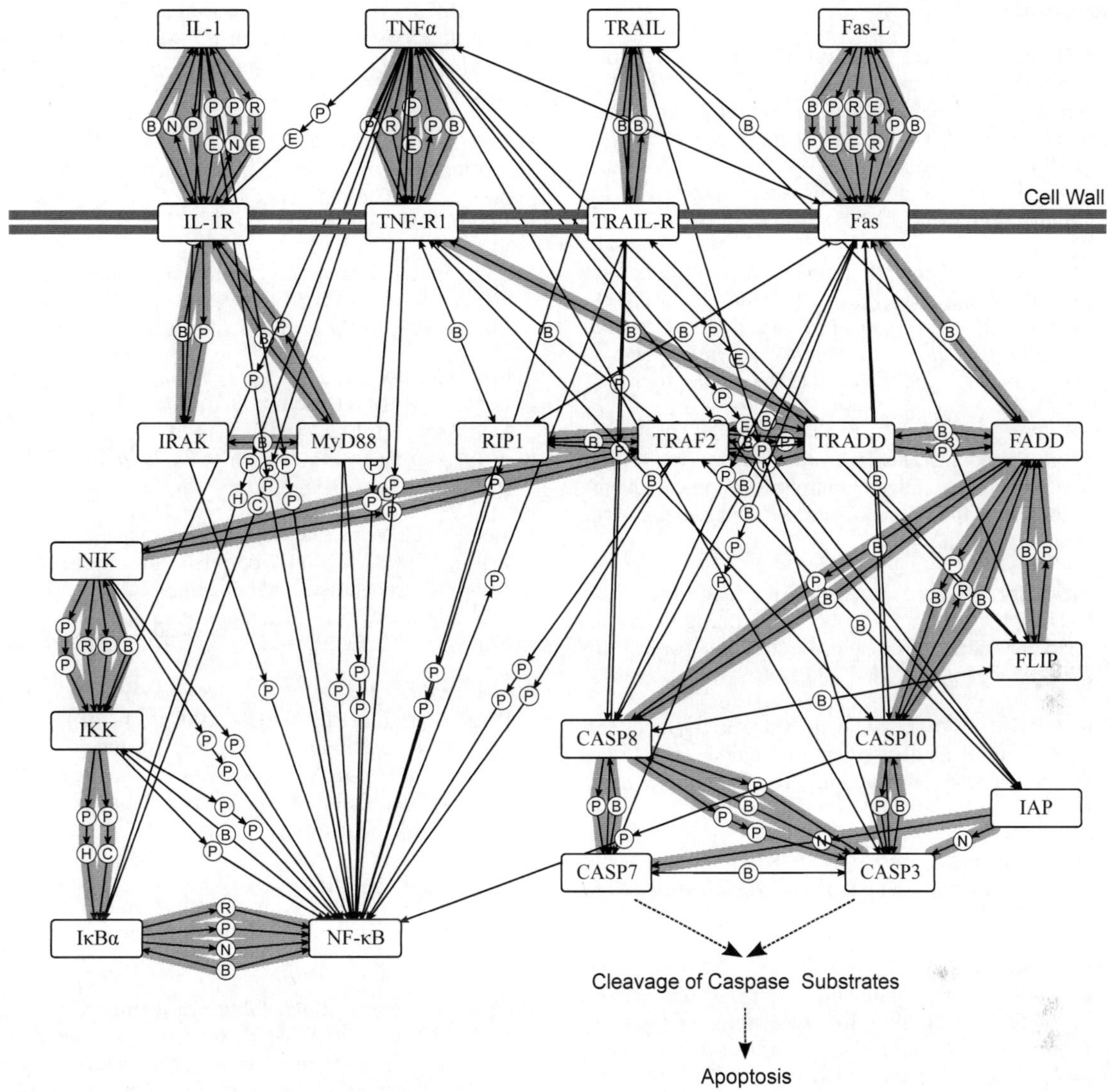

Figure 3: Extracted apoptosis event network. Events shown in the figure are selected on their prominence in the data or correspondence to known apoptosis interactions. Events corresponding to KEGG apoptosis pathway interaction partners are highlighted with a light grey background. The event types are (P)ositive regulation, (N)egative regulation, (R)egulation, gene (E)xpression, (B)inding, p(H)osphorylation, (L)ocalization and protein (C)atabolism.

presenting the information in a biologically meaningful manner.

The introduced dataset is freely available for research purposes at `http://bionlp.utu.fi/`.

Acknowledgments

This work was supported by the Academy of Finland and by Grant-in-Aid for Specially Promoted Research (MEXT, Japan). Computational resources were provided by *CSC – IT Center for Science, Ltd.*, a joint computing center for Finnish academia and industry. We thank Robert Leaman for advance access and assistance with the newest release of BANNER.

References

Jari Björne, Juho Heimonen, Filip Ginter, Antti Airola, Tapio Pahikkala, and Tapio Salakoski. 2009. Extracting complex biological events with rich graph-based feature sets. In *Proceedings of the BioNLP 2009 Workshop Companion Volume for Shared Task*, pages 10–18, Boulder, Colorado. Association for Computational Linguistics.

Jari Björne, Filip Ginter, Sampo Pyysalo, Jun'ichi Tsujii, and Tapio Salakoski. 2010a. Complex event extraction at PubMed scale. In *Proceedings of the 18th Annual International Conference on Intelligent Systems for Molecular Biology (ISMB 2010)*. In press.

Jari Björne, Juho Heimonen, Filip Ginter, Antti Airola, Tapio Pahikkala, and Tapio Salakoski. 2010b. Extracting contextualized complex biological events with rich graph-based feature sets. *Computational Intelligence*. In press.

Marie-Catherine de Marneffe, Bill MacCartney, and Christopher Manning. 2006. Generating typed dependency parses from phrase structure parses. In *Proceedings of LREC-06*, pages 449–454.

Renata Kabiljo, Andrew Clegg, and Adrian Shepherd. 2009. A realistic assessment of methods for extracting gene/protein interactions from free text. *BMC Bioinformatics*, 10(1):233.

M. Kanehisa and S. Goto. 2000. KEGG: kyoto encyclopedia of genes and genomes. *Nucleic Acids Res.*, 28:27–30, Jan.

Jun'ichi Kazama and Jun'ichi Tsujii. 2003. Evaluation and extension of maximum entropy models with inequality constraints. In *Proceedings of the 2003 Conference on Empirical Methods in Natural Language Processing*, pages 137–144.

Jin-Dong Kim, Tomoko Ohta, Sampo Pyysalo, Yoshinobu Kano, and Jun'ichi Tsujii. 2009. Overview of BioNLP'09 shared task on event extraction. In *Proceedings of the BioNLP 2009 Workshop Companion Volume for Shared Task*, pages 1–9, Boulder, Colorado. ACL.

Martin Krallinger, Florian Leitner, Carlos Rodriguez-Penagos, and Alfonso Valencia. 2008. Overview of the protein-protein interaction annotation extraction task of BioCreative II. *Genome Biology*, 9(Suppl 2):S4.

R. Leaman and G. Gonzalez. 2008. BANNER: an executable survey of advances in biomedical named entity recognition. *Pacific Symposium on Biocomputing*, pages 652–663.

David McClosky and Eugene Charniak. 2008. Self-Training for Biomedical Parsing. In *Proceedings of the 46th Annual Meeting of the Association for Computational Linguistics - Human Language Technologies (ACL-HLT'08)*, pages 101–104.

David McClosky. 2009. *Any Domain Parsing: Automatic Domain Adaptation for Natural Language Parsing*. Ph.D. thesis, Department of Computer Science, Brown University.

Makoto Miwa, Rune Sætre, Jin-Dong Kim, and Jun'ichi Tsujii. 2010. Event Extraction With Complex Event Classification Using Rich Features. *J Bioinform Comput Biol*, 8:131–146.

Alexander A. Morgan and Lynette Hirschman. 2007. Overview of BioCreative II gene normalization. In *Proceedings of BioCreative II*, pages 101–103.

Claire Nédellec. 2005. Learning Language in Logic - Genic Interaction Extraction Challenge. In J. Cussens and C. Nédellec, editors, *Proceedings of the 4th Learning Language in Logic Workshop (LLL05)*, pages 31–37.

Tomoko Ohta, Yuka Tateisi, Hideki Mima, and Jun'ichi Tsujii. 2002. GENIA corpus: An annotated research abstract corpus in molecular biology domain. In *Proceedings of the Human Language Technology Conference (HLT'02)*, pages 73–77.

Tomoko Ohta, Jin-Dong Kim, Sampo Pyysalo, Yue Wang, and Jun'ichi Tsujii. 2009a. Incorporating genetag-style annotation to genia corpus. In *Proceedings of the BioNLP 2009 Workshop*, pages 106–107, Boulder, Colorado, June. Association for Computational Linguistics.

Tomoko Ohta, Sampo Pyysalo, Kim Jin-Dong, and Jun'ichi Tsujii. 2009b. A re-evaluation of biomedical named entity - term relations. In *Proceedings of LBM'09*.

Sampo Pyysalo, Tomoko Ohta, Jin-Dong Kim, and Jun'ichi Tsujii. 2009. Static relations: a piece in the biomedical information extraction puzzle. In *Proceedings of the BioNLP 2009 Workshop*, pages 1–9, Boulder, Colorado, June. Association for Computational Linguistics.

Lorraine Tanabe, Natalie Xie, Lynne H Thom, Wayne Matten, and W John Wilbur. 2005. GENETAG: A tagged corpus for gene/protein named entity recognition. *BMC Bioinformatics*, 6(Suppl. 1):S3.

Yuka Tateisi, Akane Yakushiji, Tomoko Ohta, and Jun'ichi Tsujii. 2005. Syntax Annotation for the GENIA corpus. In *Proceedings of the IJCNLP 2005, Companion volume*, pages 222–227.

The Gene Ontology Consortium. 2000. Gene ontology: tool for the unification of biology. *Nature genetics*, 25:25–29.

A Comparative Study of Syntactic Parsers for Event Extraction

Makoto Miwa[1] **Sampo Pyysalo**[1] **Tadayoshi Hara**[1] **Jun'ichi Tsujii**[1,2,3]
[1]Department of Computer Science, the University of Tokyo, Japan
Hongo 7-3-1, Bunkyo-ku, Tokyo, Japan.
[2]School of Computer Science, University of Manchester, UK
[3]National Center for Text Mining, UK
{mmiwa,smp,harasan,tsujii}@is.s.u-tokyo.ac.jp

Abstract

The extraction of bio-molecular events from text is an important task for a number of domain applications such as pathway construction. Several syntactic parsers have been used in Biomedical Natural Language Processing (BioNLP) applications, and the BioNLP 2009 Shared Task results suggest that incorporation of syntactic analysis is important to achieving state-of-the-art performance. Direct comparison of parsers is complicated by to differences in the such as the division between phrase structure- and dependency-based analyses and the variety of output formats, structures and representations applied. In this paper, we present a task-oriented comparison of five parsers, measuring their contribution to bio-molecular event extraction using a state-of-the-art event extraction system. The results show that the parsers with domain models using dependency formats provide very similar performance, and that an ensemble of different parsers in different formats can improve the event extraction system.

1 Introduction

Bio-molecular events are useful for modeling and understanding biological systems, and their automatic extraction from text is one of the key tasks in Biomedical Natural Language Processing (BioNLP). In the BioNLP 2009 Shared Task on event extraction, participants constructed event extraction systems using a variety of different parsers, and the results indicated that the use of a parser was correlated with high ranking in the task (Kim et al., 2009). By contrast, the results did not indicate a clear preference for a particular parser, and there has so far been no direct comparison of different parsers for event extraction.

While the outputs of parsers applying the same out format can be compared using a gold standard corpus, it is difficult to perform meaningful comparison of parsers applying different frameworks. Additionally, it is still an open question to what extent high performance on a gold standard treebank correlates with usefulness at practical tasks. Task-based comparisons of parsers provide not only a way to asses parsers across frameworks but also a necessary measure of their practical applicability.

In this paper, five different parsers are compared on the bio-molecular event extraction task defined in the BioNLP 2009 Shared Task using a state-of-the-art event extraction system. The data sets share abstracts with GENIA treebank, and the treebank is used as an evaluation standard. The outputs of the parsers are converted into two dependency formats with the help of existing conversion methods, and the outputs are compared in the two dependency formats. The evaluation results show that different syntactic parsers with domain models in the same dependency format achieve closely similar performance, and that an ensemble of different syntactic parsers in different formats can improve the performance of an event extraction system.

2 Bio-molecular Event Extraction with Several Syntactic Parsers

This paper focuses on the comparison of several syntactic parsers on a bio-molecular event extraction task with a state-of-the-art event extraction system. This section explains the details of the comparison. Section 2.1 presents the event ex-

Proceedings of the 2010 Workshop on Biomedical Natural Language Processing, ACL 2010, pages 37–45,
Uppsala, Sweden, 15 July 2010.

traction task setting, following that of the BioNLP 2009 Shared Task. Section 2.2 then summarizes the five syntactic parsers and three formats adopted for the comparison. Section 2.3 described how the state-of-the-art event extraction system of Miwa et al. (2010) is modified and used for the comparison.

2.1 Bio-molecular Event Extraction

The bio-molecular event extraction task considered in this study is that defined in the BioNLP 2009 Shared Task (Kim et al., 2009)[1]. The shared task provided common and consistent task definitions, data sets for training and evaluation, and evaluation criteria. The shared task consists of three subtasks: core event extraction (Task 1), augmenting events with secondary arguments (Task 2), and the recognition of speculation and negation of the events (Task 3) (Kim et al., 2009). In this paper we consider Task 1 and Task 2. The shared task defined nine event types, which can be divided into five simple events (Gene_expression, Transcription, Protein_catabolism, Phosphorylation, and Localization) that take one core argument, a multi-participant binding event (Binding), and three regulation events (Regulation, Positive_regulation, and Negative_regulation) that can take other events as arguments.

In the two tasks considered, events are represented with a textual trigger, type, and arguments, where the trigger is a span of text that states the event in text. In Task 1 the event arguments that need to be extracted are restricted to the core arguments Theme and Cause, and secondary arguments (locations and sites) need to be attached in Task 2.

2.2 Parsers and Formats

Five parsers and three formats are adopted for the evaluation. The parsers are GDep (Sagae and Tsujii, 2007)[2], the Bikel parser (Bikel) (Bikel, 2004)[3], the Charniak-Johnson reranking parser, using David McClosky's self-trained biomedical parsing model (MC) (McClosky, 2009)[4], the C&C CCG parser, adapted to biomedical text

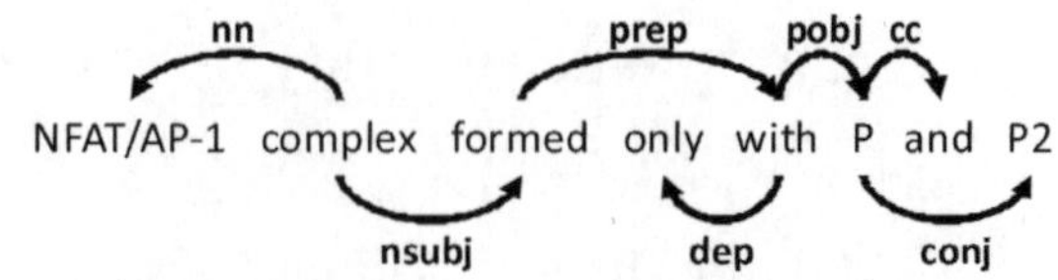

Figure 1: Stanford basic dependency tree

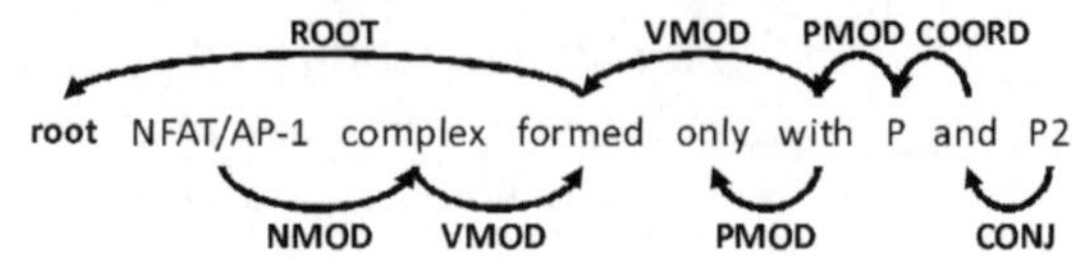

Figure 2: CoNLL-X dependency tree

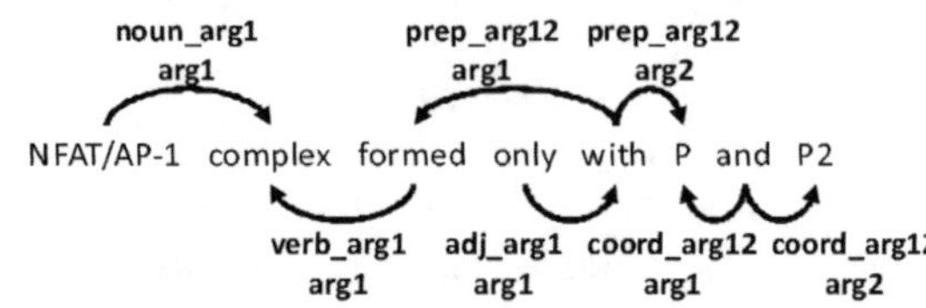

Figure 3: Predicate Argument Structure

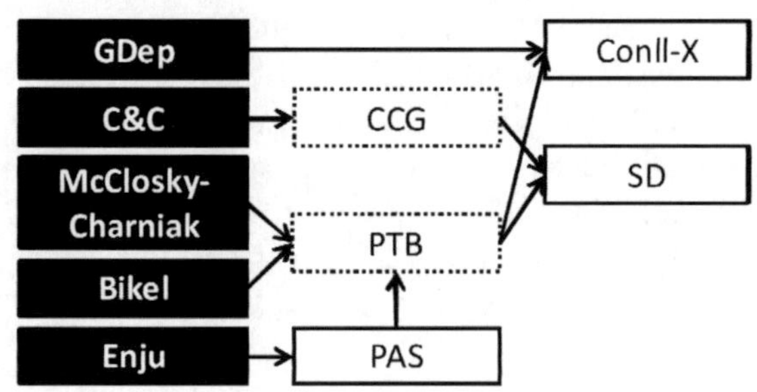

Figure 4: Format conversion dependencies in five parsers. Formats adopted for the evaluation is shown in solid boxes. SD: Stanford Dependency format, CCG: Combinatory Categorial Grammar output format, PTB: Penn Treebank format, and PAS: Predicate Argument Structure in Enju format.

(C&C) (Rimell and Clark, 2009)[5], and the Enju parser with the GENIA model (Miyao et al., 2009)[6]. The formats are Stanford Dependencies (SD) (Figure 1), the CoNLL-X dependency format (Figure 2) and the predicate-argument structure (PAS) format used by Enju (Figure 3). With the exception of Enju, the analyses of these parsers were provided by the BioNLP 2009 Shared Task organizers. Analysis of system features in the task found that the use of parser output with one of

[1]http://www-tsujii.is.s.u-tokyo.ac.jp/GENIA/SharedTask/

[2]http://www.cs.cmu.edu/~sagae/parser/gdep/

[3]http://www.cis.upenn.edu/~dbikel/software.html

[4]http://www.cs.brown.edu/~dmcc/biomedical.html

[5]http://svn.ask.it.usyd.edu.au/trac/candc/

[6]http://www-tsujii.is.s.u-tokyo.ac.jp/enju/

the formats considered here correlated with high rank at the task (Kim et al., 2009). A number of these parsers have also been shown to be effective for protein-protein interactions extraction (Miyao et al., 2009).

The five parsers operate in a number of different frameworks, reflected in their analyses. GDep is a native dependency parser that produces CoNLL-X-format dependency trees. MC and Bikel are phrase-structure parsers, and they produce Penn Treebank (PTB) format analyses. C&C is a deep parser based on Combinatory Categorial Grammar (CCG), and its native output is in a CCG-specific format. The output of C&C is converted into SD by a rule-based conversion script (Rimell and Clark, 2009). Enju is deep parser based on Head-driven Phrase Structure Grammar (HPSG) and produces a format containing predicate argument structures (PAS) along with a phrase structure tree in Enju format.

To study the contribution of the formats in which the five parsers output their analyses to task performance, we apply a number of conversions between the outputs, shown in Figure 4. The Enju PAS output is converted into Penn Treebank format using the method introduced by (Miyao et al., 2009). SD is generated from PTB by the Stanford tools (de Marneffe et al., 2006)[7], and CoNLL-X dependencies are generated from PTB by using Treebank Converter (Johansson and Nugues, 2007)[8]. We note that all of these conversions can introduce some errors in the conversion process.

With the exception of Bikel, all the applied parsers have models specifically adapted for biomedical text. Further, all of the biomedical domain models have been created with reference and for many parsers with direct training on the data of (a subset of) the GENIA treebank (Tateisi et al., 2005). The results of parsing with these models as provided for the BioNLP Shared Task are used in this comparison. However, we note that the shared task data, drawn from the GENIA event corpus (Kim et al., 2008), contains abstracts that are also in the GENIA treebank. This implies that the parsers are likely to perform better on the texts used in the shared task than on other biomedical domain text, and similarly that systems building on their output are expected to achieve best performance on this data. However, it does not invalidate comparison within the dataset. We further note that the models do not incorporate any knowledge of the event annotations of the shared task.

2.3 Event Extraction System

The system by Miwa et al. (2010) is adopted for the evaluation. The system was originally developed for finding core events (Task 1 in the BioNLP 2009 Shared Task) using Enju and GDep with the native output of these parsers. The system consists of three supervised classification-based modules: a trigger detector, an event edge detector, and a complex event detector. The trigger detector classifies each word into the appropriate event types, the event edge detector classifies each edge between an event and a protein into an argument type, and the complex event detector classifies event candidates constructed by all edge combinations, deciding between event and non-event. The system uses one-vs-all support vector machines (SVMs) for the classifications.

The system operates on one sentence at a time, building features for classification based on the syntactic analyses for the sentence provided by the two parsers as well as the sequence of the words in the sentence, including the target candidate. The features include the constituents/words around entities (triggers and proteins), the dependencies, and the shortest paths among the entities. The feature generation is format-independent regarding the shared properties of different formats, but makes use also of format-specific information when available for extracting features, including the dependency tags, word-related information (e.g. a lexical entry in Enju format), and the constituents and their head information.

The previously introduced base system is here improved with two modifications. One modification is removing two classes of features from the original features (for details of the original feature representation, we refer to (Miwa et al., 2010)); specifically the features representing governor-dependent relationships from the target word, and the features representing each event edges in the complex event detector are removed. The other modification is to use head words in a trigger expression as a gold trigger word. This modification is inspired by the part-of-speech (POS) based selection proposed by Kilicoglu and Bergler (2009).

[7] `http://www-nlp.stanford.edu/software/lex-parser.shtml`

[8] `http://nlp.cs.lth.se/software/treebank_converter/`

The system uses a head word "in" as a trigger word in a trigger expression "in the presence of" instead of using all the words of the expression. In cases where there is no head word information in a parser output, head words are selected heuristically: if a word does not modify another word in the trigger expression, the word is selected as a head word.

The system is also modified to find secondary arguments (Task 2 in the BioNLP 2009 Shared Task). The second arguments are treated as additional arguments in Task 1: the trigger detector finds secondary argument candidates, the event edge detector finds secondary argument edge candidates, and the complex event detector finds events including secondary arguments. The features are extracted using the same feature extraction method as for regulation events taking proteins as arguments.

3 Evaluation Setting

Event extraction performance is evaluated using the evaluation script provided by the BioNLP'09 shared task organizers[9] for the development data set, and the online evaluation system of the task[10] for the test data set. Results are reported under the official evaluation criterion of the task, i.e. the "Approximate Span Matching/Approximate Recursive Matching" criterion. Task 1 and Task 2 are solved at once for the evaluation.

As discussed in Section 2.2, the texts of the GENIA treebank are shared with the shared task data sets, which allows the gold annotations of the treebank to be used for reference. The GENIA treebank is converted into the Enju format with Enju. When the trees in the treebank cannot be converted into the Enju format, parse results are used instead. The GENIA treebank is also converted into PTB format[11]. The treebank is then converted into the dependency formats with the conversions described in Section 2.2. While based on manually annotated gold data, the converted treebanks are not always correct due to conversion errors.

The event extraction system described in Section 2.3 is used with the default settings shown in (Miwa et al., 2010). The positive and negative examples are balanced by placing more weight on the positive examples. The examples predicted with confidence greater than 0.5, as well as the examples with the most confident labels, are extracted. The C-values of SVMs are set to 1.0.

Some of the parse results do not include word base forms or part-of-speech (POS) tags, which are required by the event extraction system. To apply these parsers, the GENIA Tagger (Tsuruoka et al., 2005) output is adopted to add this information to the results.

	BD	CD	CDP	CTD
Task 1	**55.60**	54.35	54.59	54.42
Task 2	**53.94**	52.65	52.88	52.76

Table 1: Comparison of the F-score results with different Stanford dependency variants on the development data set with the MC parser. Results for basic dependencies (BD), collapsed dependencies (CD), collapsed dependencies with propagation of conjunct dependencies (CDP), and collapsed tree dependencies (CTD) are shown. The best score in each task is shown in bold.

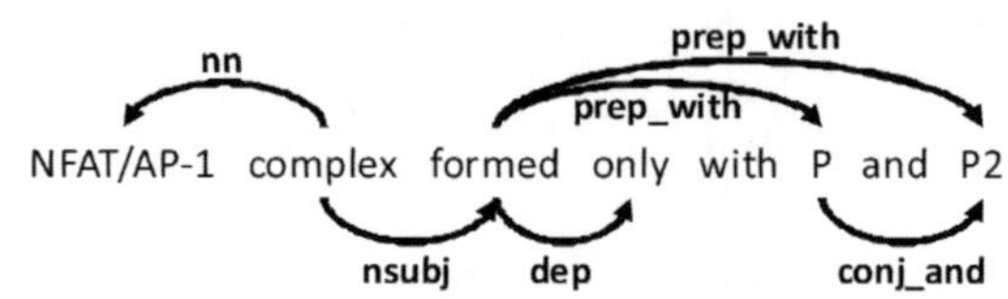

Figure 5: Stanford collapsed dependencies with propagation of conjunct dependencies

4 Evaluation

Results of event extraction with the setting in Section 2.3 will be presented in this section. Section 4.1 considers the effect of different variants of the Stanford Dependency representation. Section 4.2 presents the results of experiments with different parsers, and Section 4.3 shows the performance with ensembles of multiple parsers. Finally, the performance of the event extraction system is discussed in context of other proposed methods for the task in Section 4.4.

4.1 Stanford Dependency Setting

Stanford dependencies have four different variants: basic dependencies (BD), collapsed dependencies (CD), collapsed dependencies with propagation of conjunct dependencies (CDP), and collapsed tree dependencies (CTD) (de Marneffe and

[9] http://www-tsujii.is.s.u-tokyo.ac.jp/GENIA/SharedTask/downloads.shtml

[10] http://www-tsujii.is.s.u-tokyo.ac.jp/GENIA/SharedTask/eval-test.shtml

[11] http://categorizer.tmit.bme.hu/~illes/genia_ptb/

	BD	CD	CDP	CTD
Task 1	54.22	**54.37**	53.88	53.84
	(-1.38)	(+0.02)	(-0.71)	(-0.58)
Task 2	52.73	**52.80**	52.31	52.35
	(-1.21)	(+0.15)	(-0.57)	(-0.41)

Table 2: Comparison of the F-score results with different Stanford dependency variants without dependency types.

Manning, 2008). Except for BD, these variants do not necessarily connect all the words in the sentence, and CD and CDP do not necessarily form a tree structure. Figure 5 shows an example of CDP converted from the tree in Figure 1. To select a suitable alternative for the comparative experiments, we first compared these variants as a preliminary experiment. Table 1 shows the comparison results with the MC parser. Dependencies are generalized by removing expressions after "_" of the dependencies (e.g. "_with" in prep_with) for better performance. We find that basic dependencies give the best performance to event extraction, with little difference between the other variants. This result is surprising, as variants other than basic have features such as the resolution of conjunctions that are specifically designed for practical applications. However, basic dependendencies were found to consistently provide best performance also for the other parsers[12].

The SD variants differ from each other in two key aspects: the dependency structure and the dependency types. To gain insight into why the basic dependencies should provide better performance than other variants, we performed an experiment attempting to isolate these factors by repeating the evaluation while eliminating the dependency types. The results of this evaluation are shown in Table 2. The results indicate that the contribution of the dependency types to extraction performance differs between the variants: the expected performance drop is most notable for the basic dependencies, and for the collapsed dependencies there is even a minute increase in performance, making results for collapsed dependencies best of the untyped results (by a very narrow margin). While this result doesn't unambiguously point to a specific explanation for why basic dependencies provide best performance when types are not removed, possible explanations include errors in typing or sparseness issues causing problems in generalization for the types of non-basic dependencies. While achieving a clear resolution of the results of the comparison between SD variants requires more analysis, from a performance optimization perspective the results present an uncomplicated choice. Thus, in the following evaluation, the basic dependencies are adopted for all SD results.

4.2 Parser Comparison

Results with different parsers and different formats on the development data set are summarized in Table 3. Baseline results are produced by removing dependency (or PAS) information from the parse results. The baseline results differ between the represetations as the word base forms and POS tags produced by the GENIA tagger for use with the Stanford dependency and CoNLL-X formats are different from those for Enju, and because head word information in Enju format is used. The evaluation finds best results for both tasks with Enju, using its native output format. However, as discussed in Section 2.3, the treatment of the Enju format and the other two formats are slightly different, this result does not necessarily indicate that the Enju format is the best alternative for event extraction.

Unsurprisingly, we find that the Bikel parser, the only one in the comparison lacking a model adapted to the biomedical domain, performs worse than the other parsers. For SD, we find best results for C&C, which is notable as the parser output is processed into SD by a custom conversion, while MC output uses the *de facto* conversion of the Stanford tools. Similarly, MC produces the best result for the CoNLL-X format, which is the native output format of GDep. Enju and GDep produces comparable results to the best formats for both tasks. Overall, we find that event extraction results for the parsers applying GENIA treebank models are largely comparable for the dependency formats (SD and CoNLL-X).

The results with the data derived from the GENIA treebank can be considered as upper bounds for the parsers and formats at the task, although conversion errors are expected to lower these bounds to some extent. Even though trained on the treebank, using the parsers does not provide performance as high as that for using the GE-

[12]Collapsed tree dependencies are not evaluated on the C&C parser since the conversion is not provided.

	Task 1			Task 2		
	SD	CoNLL	PAS	SD	CoNLL	PAS
Baseline	51.05	-	50.42	49.17	-	48.88
GDep	-	55.70	-	-	54.37	-
Bikel	53.29	53.22	-	51.40	51.27	-
MC	55.60	<u>56.01</u>	-	53.94	<u>54.51</u>	-
C&C	<u>56.09</u>	-	-	<u>54.27</u>	-	-
Enju	55.48	55.74	**<u>56.57</u>**	54.06	54.37	**<u>55.31</u>**
GENIA	56.34	56.09	57.94	55.04	54.57	56.40

Table 3: Comparison of F-score results with five parsers in three different formats on the development data set. SD: Stanford basic Dependency format, CoNLL: CoNLL-X format, and PAS: Predicate Argument Structure in Enju format. Results without dependency (or PAS) information are shown as baselines. The results with the GENIA treebank (converted into PTB format and Enju format) are shown for comparison (GENIA). The best score in each task is shown in bold, and the best score in each task and format is underlined.

	Task 1			Task 2		
	C&C SD	MC CoNLL	Enju CoNLL	C&C SD	MC CoNLL	Enju CoNLL
MC CoNLL	57.44 (+1.35)	-	-	55.75 (+1.24)	-	-
Enju CoNLL	56.47 (+0.38)	56.24 (+0.23)	-	54.85 (+0.48)	54.70 (+0.19)	-
Enju PAS	57.20 (+0.63)	**57.78** (+1.21)	56.59 (+0.02)	55.75 (+0.44)	**56.39** (+1.08)	55.12 (-0.19)

Table 4: Comparison of the F-score results with parser ensembles on the development data set. C&C with Stanford basic Dependency format, MC with CoNLL-X format, Enju with CoNLL-X format, and Enju with Predicate Argument Structure in Enju format are used for the parser ensemble. The changes from single-parser results are shown in parentheses. The best score in each task is shown in bold.

NIA treebank, but in many cases results with the parsers are only slightly worse than results with the treebank. The results suggest that there is relative little remaining benefit to be gained for event extraction from improving parser performance. This supports the claim that most of the errors in event extraction are not caused by the parse errors in (Miwa et al., 2010). Experiments using the CoNLL-X format produce slightly worse results than for SD with the gold treebank data, which is at variance with the indication from parser-based results with MC and Enju. Thus, the results do not provide any systematic indication suggesting that one dependency format would be superior to the other in use for event extraction.

4.3 Event Extraction with Parser Ensemble

The four parser outputs were selected for the evaluation of a parser ensemble: C&C with Stanford basic Dependency format, MC with CoNLL-X format, Enju with CoNLL-X format, and Enju with Predicate Argument Structure in Enju format. Table 4 summarizes the parser ensemble results. We find that all ensembles of different parsers in different formats produce better results than those for single parser outputs (Table 3); by contrast, the results indicate that ensembles of the same formats (MC + Enju in CoNLL-X format) or parsers (Enju in CoNLL-X and Enju formats) produce relatively small improvements, may in some cases even reduce performance. The results thus indicate that while a parser ensemble can be effective but that it is important to apply different parsers in different formats.

Table 5 shows detailed results with three parsers with three different formats. The ensembles systematically improve F-scores in regulation and the overall performance ("All"), but the ensembles can degrade the performance for simple and binding events. Different parser outputs are shown to have their strengths and weaknesses in different event groups. The use of Enju, for exam-

	Simple	Binding	Regulation	All
		Task 1		
BL-E	75.85 / 71.09 / 73.39	40.32 / 38.17 / 39.22	30.65 / 48.16 / 37.46	46.12 / 55.60 / 50.42
BL-G	76.03 / 73.48 / 74.73	40.32 / 38.17 / 39.22	33.50 / 45.95 / 38.75	47.74 / 54.86 / 51.05
C	78.89 / 78.43 / 78.66	48.79 / 43.37 / 45.92	37.17 / 54.07 / 44.06	51.82 / 61.12 / 56.09
M	79.79 / 77.12 / 78.43	43.95 / 41.13 / 42.50	39.41 / 52.94 / 45.18	52.66 / 59.82 / 56.01
E	79.79 / 76.07 / 77.88	45.16 / 43.75 / 44.44	40.12 / 53.68 / 45.92	53.21 / 60.38 / 56.57
C+M	**80.50 / 79.05 / 79.77**	48.39 / 42.25 / 45.11	41.85 / 53.17 / 46.84	54.84 / 60.31 / 57.44
C+E	79.79 / 76.46 / 78.09	47.98 / **45.59** / 46.76	41.04 / 53.66 / 46.51	54.11 / 60.66 / 57.20
E+M	**80.50** / 77.15 / 78.79	44.35 / 42.97 / 43.65	42.26 / **55.63 / 48.03**	54.50 / **61.49** / 57.78
C+E+M	80.14 / 77.07 / 78.58	**51.61** / 42.95 / **46.89**	**42.46** / 54.30 / 47.66	**55.51** / 60.27 / **57.79**
		Task 2		
BL-E	74.60 / 69.10 / 71.75	36.55 / 34.73 / 35.62	29.89 / 47.20 / 36.60	44.74 / 53.86 / 48.88
BL-G	74.42 / 71.31 / 72.83	36.55 / 33.33 / 34.87	32.52 / 44.83 / 37.70	46.13 / 52.64 / 49.17
C	77.64 / 76.77 / 77.20	43.78 / 38.79 / 41.13	36.17 / 52.89 / 42.96	50.14 / 59.14 / 54.27
M	78.71 / 75.95 / 77.31	39.36 / 36.57 / 37.91	38.70 / 52.12 / 44.42	51.25 / 58.21 / 54.51
E	79.07 / 75.26 / 77.12	41.37 / 40.08 / 40.71	39.31 / 52.86 / 45.09	51.98 / 59.10 / 55.31
C+M	79.61 / **78.03 / 78.81**	43.37 / 36.99 / 39.93	40.93 / 52.07 / 45.83	53.31 / 58.41 / 55.75
C+E	78.89 / 75.34 / 77.08	44.18 / **40.89 / 42.47**	40.22 / 52.86 / 45.68	52.81 / 59.04 / 55.75
E+M	**79.79** / 76.33 / 78.02	40.16 / 38.76 / 39.45	41.34 / **54.69 / 47.09**	53.15 / **60.05 / 56.39**
C+E+M	79.43 / 76.25 / 77.81	**46.18** / 37.46 / 41.37	**41.54** / 53.39 / 46.72	**53.98** / 58.45 / 56.13

Table 5: Comparison of Recall / Precision / F-score results on the development data set. C&C with Stanford basic Dependency format (C), MC with CoNLL-X format (M), and Enju with Predicate Argument Structure in Enju format (E) are used for the evaluation. Results with Enju output without PAS information (BL-E) and the GENIA tagger output (BL-G) are shown as baselines. Results on simple, binding, regulation, and all events are shown. The best score in each result is shown in bold.

	Simple	Binding	Regulation	All
		Task 1		
Ours	**67.09 / 77.59 / 71.96**	49.57 / 51.65 / 50.59	**38.42 / 53.95 / 44.88**	**50.28 / 63.19 / 56.00**
Miwa	65.31 / 76.44 / 70.44	**52.16 / 53.08 / 52.62**	35.93 / 46.66 / 40.60	48.62 / 58.96 / 53.29
Björne	64.21 / 77.45 / 70.21	40.06 / 49.82 / 44.41	35.63 / 45.87 / 40.11	46.73 / 58.48 / 51.95
Riedel	N/A	23.05 / 48.19 / 31.19	26.32 / 41.81 / 32.30	36.90 / 55.59 / 44.35
		Task 2		
Ours	**65.77 / 75.29 / 70.21**	**47.56 / 49.55 / 48.54**	**38.24 / 53.57 / 44.62**	**49.48 / 61.87 / 54.99**
Riedel	N/A	22.35 / 46.99 / 30.29	25.75 / 40.75 / 31.56	35.86 / 54.08 / 43.12

Table 6: Comparison of Recall / Precision / F-score results on the test data set. MC with CoNLL-X format and Enju with Predicate Argument Structure in Enju format are used for the evaluation. Results on simple, binding, regulation, and all events are shown. Results by Miwa et al. (2010) (Miwa), Björne et al. (2009) (Björne), and Riedel et al. (2009) (Riedel) for Task 1 and Task 2 are shown for comparison. The best score in each result is shown in bold.

ple, is good for extracting regulation events, but produced weaker results for simple events. The ensembles of two parser outputs inherit both the strengths and weaknesses of the outputs in most cases, and the strengths and weaknesses of the ensembles vary depending on the combined parser outputs. The differences in performance between ensembles of the outputs of two parsers to the ensemble of the three parser outputs are +0.01 for Task 1, and 0.26 for Task 2. This result suggests that adding more different parsers does not always improve the performance. The ensemble of three parser outputs, however, shows stable performance across categories, scoring in the top two for binding, regulation, and all events, in the top four for simple events.

4.4 Performance of Event Extraction System

Table 6 shows a comparison of performance on the shared task test data. MC with CoNLL-X format and Enju with Predicate Argument Structure in Enju format are used for the evaluation, selecting one of the best performing ensemble settings in Section 4.3. The performance of the best systems in the original shared task is shown for reference ((Björne et al., 2009) in Task 1 and (Riedel et al., 2009) in Task 2). The event extraction system with our modifications performed significantly better than the best systems in the shared task, further outperforming the original system by Miwa et al. (2010). This result shows that the system applied for the comparison of syntactic parsers achieves state-of-the-art performance at event extraction. This result also shows that the system originally developed only for core events extraction can be easily extended for other arguments simply by treating the other arguments as additional arguments.

5 Related Work

Many approaches for parser comparison have been proposed in the BioNLP field. Most comparisons have used gold treebanks with intermediate formats (Clegg and Shepherd, 2007; Pyysalo et al., 2007). Application-oriented parser comparison across several formats was first introduced by Miyao et al. (2009), who compared eight parsers and five formats for the protein-protein interaction (PPI) extraction task. PPI extraction, the recognition of binary relations of between proteins, is one of the most basic information extraction tasks in the BioNLP field. Our findings do not conflict with those of Miyao et al. Event extraction can be viewed as an additional extrinsic evaluation task for syntactic parsers, providing more reliable and evaluation and a broader perspective into parser performance. An additional advantage of application-oriented evaluation on BioNLP shared task data is the availability of a manually annotated gold standard treebank, the GENIA treebank, that covers the same set of abstracts as the task data. This allows the gold treebank to be considered as an evaluation standard, in addition to comparison of performance in the primary task.

6 Conclusion

We compared five parsers and three formats on a bio-molecular event extraction task with a state-of-the-art event extraction system. The specific task considered was the BioNLP shared task, allowing the use of the GENIA treebank as a gold standard parse reference. The event extraction system, modified for a higher performance and an additional subtask, showed high performance on the shared task subtasks considered. Four of the five considered parsers were applied using biomedical models trained on the GENIA treebank, and they were found to produce similar performance. Parser ensembles were further shown to allow improvement of the performance of the event extraction system.

The contributions of this paper are 1) the comparison of several commonly used parsers on the event extraction task with a gold treebank, 2) demonstration of the usefulness of the parser ensemble on the task, and 3) the introduction of a state-of-the-art event extraction system. One limitation of this study is that the comparison between the parsers is not perfect, as the format conversions miss some information from the original formats and results with different formats depend on the ability of the event extraction system to take advantage of their strengths. To maximize comparability, the system was designed to extract features identically from similar parts of the dependency-based formats, further adding information provided by other formats, such as the lexical entries of the Enju format, from external resources. The results of this paper are expected to be useful as a guide not only for parser selection for biomedical information extraction but also for the development of event extraction systems.

The selection of compared parsers and formats in the present evaluation is somewhat limited. As future work, it would be informative to extend the comparison to other syntactic representations, such as the PTB format. Finally, the evaluation showed that the system fails to recover approximately 40% of events even when provided with manually annotated treebank data, showing that other methods and resources need to be adopted to further improve bio-molecular event extraction systems. Such improvement is left as future work.

Acknowledgments

This work was partially supported by Grant-in-Aid for Specially Promoted Research (MEXT, Japan), Genome Network Project (MEXT, Japan), and Scientific Research (C) (General) (MEXT, Japan).

References

Daniel M. Bikel. 2004. A distributional analysis of a lexicalized statistical parsing model. In *In EMNLP*, pages 182–189.

Jari Björne, Juho Heimonen, Filip Ginter, Antti Airola, Tapio Pahikkala, and Tapio Salakoski. 2009. Extracting complex biological events with rich graph-based feature sets. In *Proceedings of the BioNLP'09 Shared Task on Event Extraction*, pages 10–18.

Andrew B. Clegg and Adrian J. Shepherd. 2007. Benchmarking natural-language parsers for biological applications using dependency graphs. *BMC Bioinformatics*, 8.

Marie-Catherine de Marneffe and Christopher D. Manning. 2008. Stanford typed dependencies manual. Technical report, September.

Marie-Catherine de Marneffe, Bill MacCartney, and Christopher D. Manning. 2006. Generating typed dependency parses from phrase structure parses. In *Proceedings of the IEEE / ACL 2006 Workshop on Spoken Language Technology*.

Richard Johansson and Pierre Nugues. 2007. Extended constituent-to-dependency conversion for English. In *Proceedings of NODALIDA 2007*, Tartu, Estonia, May 25-26.

Halil Kilicoglu and Sabine Bergler. 2009. Syntactic dependency based heuristics for biological event extraction. In *Proceedings of the BioNLP 2009 Workshop Companion Volume for Shared Task*, pages 119–127, Boulder, Colorado, June. Association for Computational Linguistics.

Jin-Dong Kim, Tomoko Ohta, and Jun'ichi Tsujii. 2008. Corpus annotation for mining biomedical events from literature. *BMC Bioinformatics*, 9:10.

Jin-Dong Kim, Tomoko Ohta, Sampo Pyysalo, Yoshinobu Kano, and Jun'ichi Tsujii. 2009. Overview of bionlp'09 shared task on event extraction. In *BioNLP '09: Proceedings of the Workshop on BioNLP*, pages 1–9.

David McClosky. 2009. *Any Domain Parsing: Automatic Domain Adaptation for Natural Language Parsing*. Ph.D. thesis, Department of Computer Science, Brown University.

Makoto Miwa, Rune Sætre, Jin-Dong Kim, and Jun'ichi Tsujii. 2010. Event extraction with complex event classification using rich features. *Journal of Bioinformatics and Computational Biology (JBCB)*, 8(1):131–146, February.

Yusuke Miyao, Kenji Sagae, Rune Sætre, Takuya Matsuzaki, and Jun ichi Tsujii. 2009. Evaluating contributions of natural language parsers to protein-protein interaction extraction. *Bioinformatics*, 25(3):394–400.

Sampo Pyysalo, Filip Ginter, Veronika Laippala, Katri Haverinen, Juho Heimonen, and Tapio Salakoski. 2007. On the unification of syntactic annotations under the stanford dependency scheme: A case study on bioinfer and genia. In *Biological, translational, and clinical language processing*, pages 25–32, Prague, Czech Republic, June. Association for Computational Linguistics.

Sebastian Riedel, Hong-Woo Chun, Toshihisa Takagi, and Jun'ichi Tsujii. 2009. A markov logic approach to bio-molecular event extraction. In *BioNLP '09: Proceedings of the Workshop on BioNLP*, pages 41–49, Morristown, NJ, USA. Association for Computational Linguistics.

Laura Rimell and Stephen Clark. 2009. Porting a lexicalized-grammar parser to the biomedical domain. *J. of Biomedical Informatics*, 42(5):852–865.

Kenji Sagae and Jun'ichi Tsujii. 2007. Dependency parsing and domain adaptation with LR models and parser ensembles. In *EMNLP-CoNLL 2007*.

Yuka Tateisi, Akane Yakushiji, Tomoko Ohta, and Junfichi Tsujii. 2005. Syntax annotation for the genia corpus. In *Proceedings of the IJCNLP 2005, Companion volume*, pages 222–227, Jeju Island, Korea, October.

Yoshimasa Tsuruoka, Yuka Tateishi, Jin-Dong Kim, Tomoko Ohta, John McNaught, Sophia Ananiadou, and Jun'ichi Tsujii. 2005. Developing a robust part-of-speech tagger for biomedical text. In Panayiotis Bozanis and Elias N. Houstis, editors, *Panhellenic Conference on Informatics*, volume 3746 of *Lecture Notes in Computer Science*, pages 382–392. Springer.

Arguments of Nominals in Semantic Interpretation of Biomedical Text

Halil Kilicoglu,[1,2] Marcelo Fiszman,[2] Graciela Rosemblat,[2]
Sean Marimpietri,[3] Thomas C. Rindflesch[2]
[1]Concordia University, Montreal, QC, Canada
[2]National Library of Medicine, Bethesda, MD, USA
[3]University of California, Berkeley, CA, USA
h_kilico@cse.concordia.ca, sean.marimpietri@gmail.com
{fiszmanm,grosemblat,trindflesch}@mail.nih.gov

Abstract

Based on linguistic generalizations, we enhanced an existing semantic processor, SemRep, for effective interpretation of a wide range of patterns used to express arguments of nominalization in clinically oriented biomedical text. Nominalizations are pervasive in the scientific literature, yet few text mining systems adequately address them, thus missing a wealth of information. We evaluated the system by assessing the algorithm independently and by determining its contribution to SemRep generally. The first evaluation demonstrated the strength of the method through an F-score of 0.646 (P=0.743, R=0.569), which is more than 20 points higher than the baseline. The second evaluation showed that overall SemRep results were increased to F-score 0.689 (P=0.745, R=0.640), approximately 25 points better than processing without nominalizations.

1 Introduction

Extracting semantic relations from text and representing them as predicate-argument structures is increasingly seen as foundational for mining the biomedical literature (Kim et al., 2008). Most research has focused on relations indicated by verbs (Wattarujeekrit et al., 2004; Kogan et al., 2005). However nominalizations, gerunds, and relational nouns also take arguments. For example, the following sentence has three nominalizations, *treatment*, *suppression*, and *lactation* (nominalized forms of the verbs *treat*, *suppress*, and *lactate*, respectively). *Agonist* is derived from *agonize*, but indicates an agent rather than an event.

*Bromocriptine, an ergot alkaloid dopamine **agonist**, is a recent common **treatment** for **suppression** of **lactation** in postpartum women.*

In promoting economy of expression, nominalizations are pervasive in scientific discourse, particularly the molecular biology sublanguage, due to the highly nested and complex biomolecular interactions described (Friedman et al., 2002). However, Cohen et al. (2008) point out that nominalizations are more difficult to process than verbs. Although a few systems deal with them, the focus is often limited in both the nominalizations recognized and the patterns used to express their arguments. Inability to interpret nominal constructions in a general way limits the effectiveness of such systems, since a wealth of knowledge is missed.

In this paper, we discuss our recent work on interpreting nominal forms and their arguments. We concentrate on nominalizations; however, the analysis also applies to other argument-taking nouns. Based on training data, we developed a set of linguistic generalizations and enhanced an existing semantic processor, SemRep, for effective interpretation of a wide range of patterns used to express arguments of nominalization in clinically oriented biomedical text. We evaluated the enhancements in two ways: by examining the ability to identify arguments of nominals independently and the effect these enhancements had on the overall quality of SemRep output.

2 Background

The theoretical linguistics literature has addressed the syntax of nominalizations (e.g. Chomsky, 1970; Grimshaw, 1990; Grimshaw and Williams, 1993), however, largely as support for theoretical argumentation, rather than detailed description of the facts. Quirk et al. (1985) concentrate on the morphological derivation of

Proceedings of the 2010 Workshop on Biomedical Natural Language Processing, ACL 2010, pages 46–54,
Uppsala, Sweden, 15 July 2010.

nominalizations from verbs. Within the context of NomBank, a project dedicated to annotation of argument structure, Meyers et al. (2004a) describe the linguistics of nominalizations, emphasizing semantic roles. However, major syntactic patterns of argument realization are also noted. Cohen et al. (2008) provide a comprehensive overview of nominalizations in biomedical text. They include a review of the relevant literature, and discuss a range of linguistic considerations, including morphological derivation, passivization, transitivity, and semantic topics (e.g. agent/instrument (*activator*) vs. action/process/state (*activation*)). Based on an analysis of the PennBioIE corpus (Kulick et al., 2004), detailed distributional results are provided on alternation patterns for several nominalizations with high frequency of occurrence in biomedical text, such as *activation* and *treatment*.

In computational linguistics, PUNDIT (Dahl et al., 1987) exploited similarities between nominalizations and related verbs. Hull and Gomez (1996) describe semantic interpretation for a limited set of nominalizations, relying on WordNet (Fellbaum, 1998) senses for restricting fillers of semantic roles. Meyers et al. (1998) present a procedure which maps syntactic and semantic information for verbs into a set of patterns for nominalizations. They use NOMLEX (MacLeod et al., 1998), a nominalization lexicon, as the basis for this transformation. More recently, the availability of the NomBank corpus (Meyers et al., 2004b) has supported supervised machine learning for nominal semantic role labeling (e.g. Pradhan et al., 2004; Jiang and Ng, 2006; Liu and Ng, 2007). In contrast, Padó et al. (2008) use unsupervised machine learning for semantic role labeling of eventive nominalizations by exploiting similarities between the argument structure of event nominalizations and corresponding verbs. Gurevich and Waterman (2009) use a large parsed corpus of Wikipedia to derive lexical models for determining the underlying argument structure of nominalizations.

Nominalizations have only recently garnered attention in biomedical language processing. GeneScene (Leroy and Chen, 2005) considers only arguments of nominalizations marked by prepositional cues. Similarly, Schuman and Bergler (2006) focus on the problem of prepositional phrase attachment. In the BioNLP'09 Shared Task on Event Extraction (Kim et al., 2009), the most frequent predicates were nominals. Several participating systems discuss techniques that accommodate nominalizations (e.g. K. B. Cohen et al., 2009; Kilicoglu and Bergler, 2009). Nominalizations have not previously been addressed in clinically oriented text.

2.1 SemRep

SemRep (Rindflesch and Fiszman, 2003) automatically extracts semantic predications (logical subject-predicate-logical object triples) from unstructured text (titles and abstracts) of MEDLINE citations. It uses domain knowledge from the Unified Medical Language System® (UMLS ®) (Bodenreider, 2004), and the interaction of this knowledge and (underspecified) syntactic structure supports a robust system. SemRep extracts a range of semantic predications relating to clinical medicine (e.g. TREATS, DIAGNOSES, ADMINISTERED_TO, PROCESS_OF, LOCATION_OF), substance interactions (INTERACTS_WITH, INHIBITS, STIMULATES), and genetic etiology of disease (ASSOCIATED_WITH, PREDISPOSES, CAUSES). For example, the program identifies the following predications from input text *MRI revealed a lacunar infarction in the left internal capsule*. Arguments are concepts from the UMLS Metathesaurus and predicates are relations from the Semantic Network.

> Magnetic Resonance Imaging DIAGNOSES Infarction, Lacunar
>
> Internal Capsule LOCATION_OF Infarction, Lacunar

Processing relies on an underspecified syntactic analysis based on the UMLS SPECIALIST Lexicon (McCray et al., 1994) and the MedPost part-of-speech tagger (Smith et al., 2004). Output includes phrase identification, and for simple noun phrases, labeling of heads and modifiers.

> [HEAD(*MRI*)] [*revealed*] [*a* MOD(*lacunar*), HEAD(*infarction*)] [*in the* MOD(*left*) MOD(*internal*), HEAD(*capsule*).]

MetaMap (Aronson and Lang, 2010) maps simple noun phrases to UMLS Metathesaurus concepts, as shown below. Associated semantic types are particularly important for subsequent processing.

> [HEAD(*MRI*){Magnetic Resonance Imaging (Diagnostic Procedure)}] [*revealed*] [*a* MOD(*lacunar*), HEAD(*infarction*) {Infarction, Lacunar(Disease or Syndrome)}] [*in the* MOD(*left*) MOD(*internal*), HEAD(*capsule*) {Internal Capsule(Body Part, Organ, or Organ Component)}.]

This structure is the basis for extracting semantic predications, which relies on several mechanisms. Indicator rules map syntactic phenomena, such as verbs, nominalizations, prepositions, and modifier-head structure in the simple noun phrase to ontological predications. Examples include:

reveal (verb) → DIAGNOSES *in* (prep) → LOCATION_OF

SemRep currently has 630 indicator rules. Ontological predications are based on a modified version of the UMLS Semantic Network and have semantic types as arguments. For example:

Diagnostic Procedure DIAGNOSES Disease or Syndrome Body Part, Organ, or Organ Component LOCATION_OF Disease or Syndrome

Construction of a semantic predication begins with the application of an indicator rule, and is then constrained by two things. Arguments must satisfy syntactic restrictions for the indicator and must have been mapped to Metathesaurus concepts that match the arguments of the ontological predication indicated. As part of this processing, several syntactic phenomena are addressed, including passivization, argument coordination, and some types of relativization. For both verb and preposition indicators, underspecified syntactic rules simply ensure that subjects are on the left and objects on the right. Enhancing SemRep for nominalizations involved extending the syntactic constraints for arguments of nominalization indicators.

3 Methods

In order to gain insight into the principles underlying expression of nominal arguments, we first determined the 50 most common nominalizations in MEDLINE citations that also occur in the UMLS SPECIALIST Lexicon, and then analyzed a corpus of 1012 sentences extracted from 476 citations containing those nominalizations. We further limited these sentences to those with nominalizations containing two overt arguments (since SemRep only extracts predications with two arguments), resulting in a final set of 383 sentences. We determined 14 alternation patterns for nominalizations based on this analysis and devised an algorithm to accommodate them. We then conducted two evaluations, one to assess the effectiveness of the algorithm independently of other considerations and another to assess the contribution of enhanced nominalization processing to SemRep generally.

3.1 Nominal Alternations

Much work in identifying arguments of nominalizations assigns semantic role, such as agent, patient, etc., but SemRep does not. In this analysis, arguments are logical subject and object. Relational nouns often allow only one argument (e.g. *the weight of the evidence*), and either one or both of the arguments of a nominalization or gerund may be left unexpressed. SemRep doesn't interpret nominalizations with unexpressed arguments. If both arguments appear, they fall into one of several patterns, and the challenge in nominalization processing is to accommodate these patterns. Cohen et al. (2008) note several such patterns, including those in which both arguments are to the right of the nominalization, cued by prepositions (*treatment of fracture with surgery*), the nominalization separates the arguments (*fracture treatment with surgery, surgical treatment for fracture*), and both arguments precede the nominalizations, as modifiers of it (*surgical fracture treatment* and *fracture surgical treatment*).

Cohen et al. (2008) do not list several patterns we observed in the clinical domain, including those in which the subject appears to the right marked by a verb (*the treatment of fracture is surgery*) or as an appositive (*the treatment of fracture, surgery*), and those in which the subject appears to the left and the nominalization is either in a prepositional phrase (*surgery in the treatment of fracture, surgery in fracture treatment*) or is preceded by a verb or is parenthetical (*surgery is (the best) treatment for fracture*; *surgery is (the best) fracture treatment; surgery, the best fracture treatment*). One pattern, in which both arguments are on the right and the subject precedes the object, is seen most commonly in the clinical domain when the nominalization has a lexically specified cue (e.g. *the contribution of stem cells to kidney repair*). The nominal alternation patterns are listed in Table 1.

Generalizations about arguments of nominalizations are based on the position of the arguments, both with respect to each other and to the nominalization, and whether they modify the nominalization or not. A modifying argument is internal to the simple noun phrase of which the nominalization is the head; other arguments (both to the left and to the right) are external. (Relativization is considered external to the simple noun phrase.)

[NOM] [PREP OBJ] [PREP SUBJ] *Treatment of fracture with surgery*
[NOM] [PREP OBJ], [SUBJ] *The treatment of fracture, surgery*
[NOM] [PREP OBJ] ([SUBJ]) *The treatment of fracture (surgery)*
[NOM] [PREP OBJ] [BE] [SUBJ] *The treatment of fracture is surgery*
[NOM] [PREP SUBJ] [PREP OBJ] *Treatment with surgery of fracture*
[SUBJ NOM] [PREP OBJ] *Surgical treatment of fracture*
[SUBJ] [PREP NOM] [PREP OBJ] *Surgery in the treatment of fracture*
[SUBJ] [BE] [NOM] [PREP OBJ] *Surgery is the treatment of fracture*
[OBJ NOM] [BE] [SUBJ] *Fracture treatment is surgery*
[OBJ NOM] [PREP SUBJ] *Fracture treatment with surgery*
[SUBJ] [PREP OBJ NOM] *Surgery for fracture treatment*
[SUBJ] [BE] [OBJ NOM] *Surgery is the fracture treatment*
[SUBJ OBJ NOM] *Surgical fracture treatment*
[OBJ SUBJ NOM] *Fracture surgical treatment*

Table 1. Patterns

Argument cuing plays a prominent role in defining these patterns. A cue is an overt syntactic element associated with an argument, and can be a preposition, a verb (most commonly a form of *be*), a comma, or parenthesis. A cued argument is in a dependency with the cue, which is itself in a dependency with the nominalization. The cue must occur between the nominalization and the argument, whether the argument is to the right (e.g. *treatment **of** fracture*) or to the left (e.g. *surgery **in** the treatment*). Prepositional cues for the objects of some nominalizations are stipulated in the lexicon; some of these are obligatory (e.g. *contribution – to*), while others are optional (*treatment – for*).

External arguments of nominalizations must be cued, and cues unambiguously signal the role of the argument, according to the following cuing rules (Cohen et al., 2008). Verbs, comma, parenthesis, and the prepositions *by, with,* and *via* cue subjects only. (*By* is used for semantic role agent and *with* for instrument, but SemRep does not exploit this distinction.) *Of* cues subjects only if the nominalization has an obligatory (object) cue; it must cue objects otherwise. There is a class of nominalizations (e.g. *cause*) that do not allow a prepositionally cued subject. Considerable variation is seen in the order of subject and object; however, if the subject intervenes between the nominalization and the object, both must have equal cuing status (the only possibilities are that both be either uncued or cued with a preposition).

3.2 Algorithm

In extending SemRep for identifying arguments of nominalizations, existing machinery was exploited, namely shallow parsing, mapping simple noun phrases to Metathesaurus concepts, and the application of indicator rules to map nominalizations to enhanced Semantic Network ontological predications (which imposes restrictions on the semantic type of arguments). Finally, syntactic argument identification was enhanced specifically for nominalizations and exploits the linguistic generalizations noted. For example in the sentence below, phrases have been identified and *cervical cancer* has been mapped to the Metathesaurus concept "Cervix carcinoma" with semantic type 'Neoplastic Process', and *vaccination* to "Vaccination" ('Therapeutic or Preventive Procedure'). An indicator rule for *prevention* maps to the ontological predication "Therapeutic or Preventive Procedure PREVENTS Neoplastic Process" (among others) in generating the predication: "Vaccination PREVENTS Cervix carcinoma."

> *Therefore, **prevention of** cervical cancer **with** HPV vaccination may have a significant financial impact.*

Processing to identify arguments for *prevention* begins by determining whether the nominalization has a lexically specified object cue. This information is needed to determine the cuing function of *of*. Since it is common for there to be at least one argument on the right, identification of arguments begins there. Arguments on the right are external and must be cued. If a cued argument is found, its role is determined by the argument cuing rules. Since *prevention* does not have a lexically specified cue, *of* marks its object. Further, the semantic type of the concept for the object of *of* matches the object of the ontological predication ('Neoplastic Process').

The algorithm next looks to the right of the first argument for the second argument. Since processing addresses only two arguments for nominalizations, subject and object, once the role of the first has been determined, the second can be inferred. For cued arguments, the process checks that the cue is compatible with the cuing rules. In all cases, the relevant semantic type must match the subject of the ontological predication. In this instance, *with* cues subjects and 'Therapeutic or Preventive Process' matches the subject of the ontological predication indicated.

If only one noun phrase to the right satisfies the argument cuing rules, the second argument must be on the left. A modifier immediately to the left of the nominalization (and thus an internal argument) is sought first, and its role inferred from the first argument. Since internal arguments are not cued, there is no need to ensure cuing compatibility. The predication "Operative Surgical Procedures TREATS Pregnancy, Ectopic" is found for *resolution* in

> *Surgical* ***resolution of*** *an ectopic pregnancy in a captive gerenuk (Litocranius walleri walleri).*

Resolution is an indicator for the ontological predication "Therapeutic or Preventive Procedure TREATS Disease or Syndrome." *Surgical* maps to "Operative Surgical Procedures" ('Therapeutic or Preventive Procedure'), which matches the subject of this predication, and *ectopic pregnancy* maps to "Pregnancy, Ectopic" ('Disease or Syndrome'), which matches its object. *Of* marks the object of *resolution*.

An argument to the left of a nominalization may be external, in which case a cue is necessary. *For* preceding *treatment* satisfies this requirement in the following sentence.

> *Preclinical data have supported the use of fludarabine and cyclophosphamide (FC) in combination* ***for*** *the* ***treatment of*** *indolent lymphoid malignancies.*

The two drugs in this sentence map to concepts with semantic type 'Pharmacologic Substance' and the malignancy has 'Neoplastic Process', as above. There is an ontological predication for TREATS with subject 'Pharmacologic Substance'. After coordination processing in SemRep, two predications are generated for *treatment*:

Cyclophosphamide TREATS Malignant lymphoid neoplasm Fludarabine TREATS Malignant lymphoid neoplasm

If there is no argument to the right, both arguments must be on the left. A modifier immediately to the left of the nominalization is sought first. Given the properties of cuing (the cue intervenes between the argument and the nominalization), if both arguments occur to the left, at least one of them must be internal, since it is not possible to have more than one external argument on the left (e.g. **Surgery is fracture for treatment*). The role of the first argument is found based on semantic type. The first modifier to the left of *treatment* in the following sentence is *epilepsy*, which has semantic type 'Disease or Syndrome', matching the object of the ontological predication for TREATS.

> *Patients with most chances of benefiting from surgical epilepsy* ***treatment***

The second modifier to the left, *surgical* maps to the concept "Operative Surgical Procedures," whose semantic type matches the subject of the ontological predication. These conditions allow construction of the predication "Operative Surgical Procedures TREATS Epilepsy."

In the next sentence, the indicator rule for *prediction* maps to the ontological predication "Amino Acid, Peptide, or Protein PREDISPOSES Disease or Syndrome."

> *The potential clinical role of measuring these apolipoproteins* ***for*** *ischemic stroke* ***prediction*** *warrants further study.*

Ischemic stroke satisfies the object of this predication and *apolipoproteins* the subject. Since the external subject is cued by *for,* all constraints are satisfied and the predication "Apolipoproteins PREDISPOSES Ischemic stroke" is generated.

3.3 Evaluation

Three-hundred sentences from 239 MEDLINE citations (titles and abstracts) were selected for annotating a test set. Some had previously been selected for various aspects of SemRep evaluation; others were chosen randomly. A small number (30) were sentences in the GENIA event corpus (Kim et al., 2008) with bio-event-triggering nominalizations. Annotation was conducted by three of the authors. One, a linguist (A), judged all sentences, while the other two, a computer scientist (B) and a medical informatics researcher (C), annotated a subset. Annotation was not limited to nominalizations. The statistics regarding the individual annotations are given below. The numbers in parentheses show the number of annotated predications indicated by nominalizations.

Annotator	# of Sentences	# of Predications
A	300	533 (286)
B	200	387 (190)
C	132	244 (134)

Table 2. Annotation statistics

As guidance, annotators were provided UMLS Metathesaurus concepts for the sentences. However, they consulted the Metathesaurus directly to check questionable mappings. Annotation focused on the 25 predicate types SemRep addresses.

We measured inter-annotator agreement, defined as the F-score of one set of annotations, when the second is taken as the gold standard. After individual annotations were complete, two annotators (A and C) assessed all three sets of annotations and created the final reference standard. The reference standard has 569 predications, 300 of which (52.7%) are indicated by nominalizations. We further measured the agreement between individual sets of annotations and the reference standard. Results are given below:

Annotator pair	# of Sentences	IAA
A-B	200	0.794
A-C	132	0.974
B-C	103	0.722
A-Gold	300	0.925
B-Gold	200	0.889
C-Gold	132	0.906

Table 3. Inter-annotator agreement

We performed two evaluations. The first (*eval1*) evaluated nominalizations in isolation, while the second (*eval2*) assessed the effect of the enhancements on overall semantic interpretation in SemRep. For *eval1*, we restricted SemRep to extract predications indicated by nominalizations only. The baseline was a nominalization argument identification rule which simply stipulates that the subject of a predicate is a concept to the left (starting from the modifier of the nominalization, if any), and the object is a concept to the right. This baseline implements the underspecification principle of SemRep, without any additional logic. We compared the results from this baseline to those from the algorithm described above to identify arguments of nominalizations. The gold standard for *eval1* was limited to predications indicated by nominalizations.

We investigated the effect of nominalization processing on SemRep generally in *eval2*, for which the baseline implementation was SemRep with no nominalization processing. The results for this baseline were evaluated against those obtained using SemRep with no restrictions. Typical evaluation metrics, precision, recall, and F-score, were calculated.

4 Results and Discussion

The results for the two evaluations are presented below.

	Precision	Recall	F-Score
eval1			
Baseline	0.484	0.359	0.412
With NOM	0.743	0.569	0.645
eval2			
Baseline	0.640	0.333	0.438
With NOM	0.745	0.640	0.689

Table 4. Evaluation results

Results illustrate the importance of nominalization processing for effectiveness of semantic interpretation and show that the SemRep methodology naturally extends to this phenomenon. With a single, simple, rule (*eval1 baseline*), SemRep achieves an F-score of 0.412. With additional processing based on linguistic generalizations, F-score improves more than 20 points. Further, the addition of nominalization processing not only enhances the coverage of SemRep (more than 30 points), but also increases precision (more than 10 points). While nominalizations are generally considered more difficult to process than verbs (Cohen et al., 2008), we were able to accommodate them with greater precision than other types of indicators, including verbs (0.743 vs. 0.64 in *eval1 with NOM* vs. *eval2 baseline*) with our patterns.

	Precision	Recall	F-Score
eval1			
Baseline	0.233	0.140	0.175
With NOM	0.690	0.400	0.506
eval2			
Baseline (No NOM)	0.667	0.278	0.392
With NOM	0.698	0.514	0.592

Table 5. Results for molecular biology sentences

Limiting the evaluation to sentences focusing on biomolecular interactions (from GENIA), while not conclusive due to the small number of sentences (30), also shows similar patterns, as shown in Table 5. As expected, while overall quality of predications is lower, since molecular biology text is significantly more complex than that in the clinical domain, improvements with nominalization processing are clearly seen.

Errors were mostly due to aspects of SemRep orthogonal to but interacting with nominalization processing. Complex coordination structure was the main source of recall errors, as in the following example.

> *RESULTS: The best* ***predictors of*** *incident metabolic syndrome* ***were*** *waist circumference (odds ratio [OR] 1.7 [1.3-2.0] per 11 cm), HDL cholesterol (0.6 [0.4-0.7] per 15 mg/dl), and proinsulin (1.7 [1.4-2.0] per 3.3 pmol/l).* [PMID 14988303]

While the system was able to identify the predication "Waist circumference PREDISPOSES Metabolic syndrome," it was unable to find the predications below, due to its inability to identify the coordination of *waist circumference*, *HDL cholesterol,* and *proinsulin.*

(FN) Proinsulin PREDISPOSES Metabolic syndrome (FN) High Density Lipoprotein Cholesterol PREDISPOSES Metabolic syndrome

Mapping of noun phrases to the correct UMLS concepts (MetaMap) is a source of both false positives and false negatives, particularly in the context of the molecular biology sentences, where acronyms and abbreviations are common and their disambiguation is nontrivial (Okazaki et al., 2010). For example, in the following sentence

> *PTK* ***inhibition with*** *Gen attenuated both LPS-induced NF-kappaB DNA binding and TNF-alpha production in human monocytes.* [PMID 10210645]

PTK was mapped to "Ephrin receptor EphA8" rather than to "Protein Tyrosine Kinase", causing both a false positive and a false negative.

(FP) Genistein INHIBITS Ephrin receptor EphA8 (FN) Genistein INHIBITS Protein Tyrosine Kinase

Some errors were due to failure to recognize a relative clause by SemRep. Only the head of such a structure is allowed to be an argument outside the structure. In the sentence below, the subject of *treatment* is *hyperthermic intraperitoneal intraoperative chemotherapy*, which is the head of the reduced relative clause, *after cytoreductive surgery*.

> *Hyperthermic intraperitoneal intraoperative chemotherapy after cytoreductive surgery* ***for the treatment of*** *abdominal sarcomatosis: clinical outcome and prognostic factors in 60 consecutive patients.* [PMID 15112276]

SemRep failed to recognize the relative clause, and therefore the nominalization algorithm took the noun phrase inside it as the subject of *treatment*, since it satisfies both semantic type and argument constraints.

(FP) Cytoreductive surgery TREATS Sarcamatosis NOS (FN) intraperitoneal therapy TREATS Sarcamatosis NOS

A small number of errors were due solely to nominalization processing. In the following sentence, the object of *contribution* is cued with *in,* rather than lexically specified *to*, which causes a recall error.

> *Using SOCS-1 knockout mice, we investigated the* ***contribution of*** *SOCS-1* ***in*** *the development of insulin resistance induced by a high-fat diet (HFD).* [PMID 18929539]

(FN) Cytokine Inducible SH-2 Containing Protein PREDISPOSES Insulin Resistance

Accurate identification of the arguments of nominalizations in the molecular biology subdomain is more challenging than in clinically-oriented text. Some of the syntactic structure responsible for this complexity is discussed by K. B. Cohen et al. (2009). In particular, they note the problem of an argument being separated from the nominalization, and point out the problem of specifying the intervening structure. Although we have not focused on molecular biology, the analysis developed for clinical medicine shows promise in that domain as well. One relevant extension could address the syntactic configuration in which intervening structure involves an argument of a nominalization shared with a verb occurring to the left of the nominalization, as *induced* and *activation* interact in the following sentence:

> *IL-2* ***induced*** *less STAT1 alpha* ***activation*** *and IFN-alpha* ***induced*** *greater STAT5* ***activation*** *in NK3.3 cells compared with preactivated primary NK cells.* [PMID 8683106]

This could be addressed with an extension of our rule that subjects of nominalizations can be cued with verbs. With respect to argument identification, *induce* can function like a form of *be.*

5 Conclusion

We discuss a linguistically principled implementation for identifying arguments of nominalizations in clinically focused biomedical text. The full range of such structures is rarely addressed by existing text mining systems, thus missing valuable information. The algorithm is implemented inside SemRep, a general semantic interpreter for biomedical text. We evaluated the system both by assessing the algorithm independently and by determining the contribution it makes to SemRep generally. The first evaluation resulted in an F-score of 0.646 (P=0.743, R=0.569), which is 20 points higher than the baseline, while the second showed that overall SemRep results were increased to F-score 0.689 (P=0.745, R=0.640), approximately 25 points better than processing without nominalizations.

Since our nominalization processing is by extending SemRep, rather than by creating a dedicated system, we provide the interpretation of these structures in a broader context. An array of semantic predications generated by mapping to an ontology (UMLS) normalizes the interpretation of verbs and nominalizations. Processing is linguistically based, and several syntactic phenomena are addressed, including passivization, argument coordination, and relativization. The benefits of such processing include effective applications for extracting information on genetic diseases from text (Masseroli et al., 2006), as well as research in medical knowledge summarization (Fiszman et al., 2004; Fiszman et al., 2009), literature-based discovery (Ahlers et al., 2007; Hristovski et al., 2010), and enhanced information retrieval (Kilicoglu et al., 2008; T. Cohen et al., 2009).

Acknowledgments

This study was supported in part by the Intramural Research Program of the National Institutes of Health, National Library of Medicine.

References

C. B. Ahlers, D. Hristovski, H. Kilicoglu, T. C. Rindflesch. 2007. Using the literature-based discovery paradigm to investigate drug mechanisms. In *Proceedings of AMIA Annual Symposium*, pages 6-10.

A. R. Aronson and F.-M. Lang. 2010. An overview of MetaMap: historical perspective and recent advances. *Journal of the American Medical Informatics Association*, 17:229-236.

O. Bodenreider. 2004. The Unified Medical Language System (UMLS): integrating biomedical terminology. *Nucleic Acids Research*, 32(Database issue):D267-70.

N. Chomsky. 1970. Remarks on nominalization. In Jacobs, Roderick, and Peter S. Rosenbaum (eds.) *Readings in English transformational grammar.* Boston: Ginn and Company, pages 184-221.

K. B. Cohen, M. Palmer, L. Hunter. 2008. Nominalization and alternations in biomedical language. *PLoS ONE,* 3(9): e3158.

K. B. Cohen, K. H. Verspoor, H. L. Johnson, C. Roeder, P. V. Ogren, W. A. Baumgartner, E. White, H. Tipney, L. Hunter. 2009. High-precision biological event extraction with a concept recognizer. In *Proceedings of the BioNLP 2009 Workshop Companion Volume for Shared Task*, pages 50-58.

T. Cohen, R. Schvaneveldt, T. C. Rindflesch. 2009. Predication-based semantic indexing: Permutations as a means to encode predications in semantic space. In *Proceedings of AMIA Annual Symposium*, pages 114-118.

D. A. Dahl, M. S. Palmer, R. J. Passonneau. 1987. Nominalizations in PUNDIT. In *Proceedings of ACL*, pages 131-139.

C. Fellbaum. 1998. *WordNet: An Electronic Lexical Database*. The MIT Press, Cambridge, MA.

M. Fiszman, D. Demner-Fushman, H. Kilicoglu, T. C. Rindflesch. 2009. Automatic summarization of MEDLINE citations for evidence-based medical treatment: A topic-oriented evaluation. *Journal of Biomedical Informatics,* 42(5):801-813.

M. Fiszman, T. C. Rindflesch, H. Kilicoglu. 2004. Abstraction summarization for managing the biomedical research literature. In *Proceedings of HLT/NAACL Workshop on Computational Lexical Semantics*, pages 76-83.

C. Friedman, P. Kra, A. Rzhetsky. 2002. Two biomedical sublanguages: a description based on the theories of Zellig Harris. *Journal of Biomedical Informatics,* 35:222–235.

J. Grimshaw. 1990. *Argument Structure.* MIT Press, Cambridge, MA.

J. Grimshaw and E. Williams. 1993. Nominalizations and predicative prepositional phrases. In J. Pustejovsky (ed.) *Semantics and the Lexicon.* Dordrecht: Kluwer Academic Publishers, pages 97-106.

O. Gurevich and S. A. Waterman. 2009. Mining of parsed data to derive deverbal argument structure. In *Proceedings of the 2009 Workshop on Grammar Engineering Across Frameworks.* pages 19-27.

D. Hristovski, A. Kastrin, B. Peterlin, T. C. Rindflesch. 2010. Combining semantic relations and DNA microarray data for novel hypothesis generation. In C. Blaschke, H. Shatkay (Eds.) *ISMB/ECCB2009, Lecture Notes in Bioinformatics,* Heidelberg: Springer-Verlag, pages 53-61.

R. D. Hull and F. Gomez. 1996. Semantic interpretation of nominalizations. In *Proceedings of AAAI*, pages 1062-1068.

Z. P. Jiang and H. T. Ng. 2006. Semantic role labeling of NomBank: A maximum entropy approach. In *Proceedings of EMNLP'06*, pages 138–145.

H. Kilicoglu and S. Bergler. 2009. Syntactic dependency based heuristics for biological event extraction. In *Proceedings of the BioNLP 2009 Workshop Companion Volume for Shared Task*, pages 119-127.

H. Kilicoglu, M. Fiszman, A. Rodriguez, D. Shin, A. M. Ripple, T. C. Rindflesch. 2008. Semantic MEDLINE: A Web application to manage the results of PubMed searches. In *Proceedings of SMBM'08,* pages 69-76.

J-D. Kim, T. Ohta, S. Pyysalo, Y. Kano, J. Tsujii. 2009. Overview of BioNLP'09 Shared Task on Event Extraction. In *Proceedings of the BioNLP 2009 Workshop Companion Volume for Shared Task*, pages 1-9.

J-D. Kim, T. Ohta, J. Tsujii. 2008. Corpus annotation for mining biomedical events from literature. *BMC Bioinformatics*, 9(1):10.

Y. Kogan, N. Collier, S. Pakhomov, M. Krauthammer. 2005. Towards semantic role labeling & IE in the medical literature. In *Proceedings of AMIA Annual Symposium,* pages 410–414.

S. Kulick, A. Bies, M. Liberman, M. Mandel, R. McDonald, M. Palmer. A. Schein, L. Ungar. 2004. Integrated annotation for biomedical information extraction. In *Proceedings of BioLINK: Linking Biological Literature, Ontologies and Databases*, pages 61–68.

G. Leroy and H. Chen. 2005. Genescene: An ontology-enhanced integration of linguistic and co-occurrence based relations in biomedical texts. *Journal of the American Society for Information Science and Technology,* 56(5): 457–468.

C. Liu and H. Ng. 2007. Learning predictive structures for semantic role labeling of NomBank. In *Proceedings of ACL*, pages 208–215.

C. Macleod, R. Grishman, A. Meyers, L. Barrett, R. Reeves. 1998. NOMLEX: A lexicon of nominalizations. In *Proceedings of EURALEX'98.*

M. Masseroli, H. Kilicoglu, F-M. Lang, T. C. Rindflesch. 2006. Argument-predicate distance as a filter for enhancing precision in extracting predications on the genetic etiology of disease. *BMC Bioinformatics,* 7:291.

A. T. McCray, S. Srinivasan, A. C. Browne. 1994. Lexical methods for managing variation in biomedical terminologies. In *Proceedings of 18th Annual Symposium on Computer Applications in Medical Care*, pages 235–239.

A. Meyers, C. Macleod, R. Yanbarger, R. Grishman, L. Barrett, R. Reeves. 1998. Using NOMLEX to produce nominalization patterns for information extraction. In *Proceedings of the Workshop on Computational Treatment of Nominals (COLING/ACL)*, pages 25-32.

A. Meyers, R. Reeves, C. Macleod, R. Szekely, V. Zielinska, B. Young, R. Grishman. 2004a. Annotating noun argument structure for NomBank. In *Proceedings of LREC.*

A. Meyers, R. Reeves, C. Macleod, R. Szekely, V. Zielinska, B. Young, R. Grishman. 2004b. The NomBank project: An interim report. In *Proceedings of HLT-NAACL 2004 Workshop: Frontiers in Corpus Annotation,* pages 24–31.

N. Okazaki, S. Ananiadou, J. Tsujii. 2010. Building a high quality sense inventory for improved abbreviation disambiguation. *Bioinformatics*: btq129+.

S. Padó, M. Pennacchiotti, C. Sporleder. 2008. Semantic role assignment for event nominalisations by leveraging verbal data. In *Proceedings of CoLing'08*, pages 665-672.

S. Pradhan, H. Sun, W. Ward, J. Martin, D. Jurafsky. 2004. Parsing arguments of nominalizations in English and Chinese. In *Proceedings of HLT/NAACL*, pages 141–144.

R. Quirk, S. Greenbaum, G. Leech, J. Svartvik. 1985. *A Comprehensive Grammar of the English Language.* Longman, London.

T. C. Rindflesch and M. Fiszman. 2003. The interaction of domain knowledge and linguistic structure in natural language processing: Interpreting hypernymic propositions in biomedical text. *Journal of Biomedical Informatics*, 36(6):462-77.

J. Schuman and S. Bergler. 2006. Postnominal prepositional phrase attachment in proteomics. In *Proceedings of BioNLP Workshop on Linking Natural Language Processing and Biology*, pages 82–89.

L. Smith, T. C. Rindflesch, W. J. Wilbur. 2004. MedPost: a part-of-speech tagger for biomedical text. *Bioinformatics*, 20(14):2320-2321.

T. Wattarujeekrit, P. K. Shah, N. Collier. 2004. PASBio: Predicate-argument structures for event extraction in molecular biology. *BMC Bioinformatics,* 5:155.

Improving Summarization of Biomedical Documents using Word Sense Disambiguation

Laura Plaza[†] lplazam@fdi.ucm.es
Mark Stevenson[*] m.stevenson@dcs.shef.ac.uk
Alberto Díaz[†] albertodiaz@fdi.ucm.es

[†] Universidad Complutense de Madrid, C/Prof. José García Santesmases, 28040 Madrid, Spain
[*] University of Sheffield, Regent Court, 211 Portobello St., Sheffield, S1 4DP, UK

Abstract

We describe a concept-based summarization system for biomedical documents and show that its performance can be improved using Word Sense Disambiguation. The system represents the documents as graphs formed from concepts and relations from the UMLS. A degree-based clustering algorithm is applied to these graphs to discover different themes or topics within the document. To create the graphs, the MetaMap program is used to map the text onto concepts in the UMLS Metathesaurus. This paper shows that applying a graph-based Word Sense Disambiguation algorithm to the output of MetaMap improves the quality of the summaries that are generated.

1 Introduction

Extractive text summarization can be defined as the process of determining salient sentences in a text. These sentences are expected to condense the relevant information regarding the main topic covered in the text. Automatic summarization of biomedical texts may benefit both health-care services and biomedical research (Reeve et al., 2007; Hunter and Cohen, 2006). Providing physicians with summaries of their patient records can help to reduce the diagnosis time. Researchers can use summaries to quickly determine whether a document is of interest without having to read it all.

Summarization systems usually work with a representation of the document consisting of information that can be directly extracted from the document itself (Erkan and Radev, 2004; Mihalcea and Tarau, 2004). However, recent studies have demonstrated the benefit of summarization based on richer representations that make use of external knowledge sources (Plaza et al., 2008; Fiszman et al., 2004). These approaches can represent semantic associations between the words and terms in the document (i.e. synonymy, hypernymy, homonymy or co-occurrence) and use this information to improve the quality of the summaries. In the biomedical domain the Unified Medical Language System (UMLS) (Nelson et al., 2002) has proved to be a useful knowledge source for summarization (Fiszman et al., 2004; Reeve et al., 2007; Plaza et al., 2008). In order to access the information contained in the UMLS, the vocabulary of the document being summarized has to be mapped onto it. However, ambiguity is common in biomedical documents (Weeber et al., 2001). For example, the string *"cold"* is associated with seven possible meanings in the UMLS Metathesuarus including *"common cold"*, *"cold sensation"* , *"cold temperature"* and *"Chronic Obstructive Airway Disease"*. The majority of summarization systems in the biomedical domain rely on MetaMap (Aronson, 2001) to map the text onto concepts from the UMLS Metathesaurus (Fiszman et al., 2004; Reeve et al., 2007). However, MetaMap frequently fails to identify a unique mapping and, as a result, various concepts with the same score are returned. For instance, for the phrase *"tissues are often cold"* MetaMap returns three equally scored concepts for the word "cold": *"common cold"*, *"cold sensation"* and *"cold temperature"*.

The purpose of this paper is to study the effect of lexical ambiguity in the knowledge source on semantic approaches to biomedical summarization. To this end, the paper describes a concept-based summarization system for biomedical documents that uses the UMLS as an external knowledge source. To address the word ambiguity problem, we have adapted an existing WSD system (Agirre and Soroa, 2009) to assign concepts from the UMLS. The system is applied to the summarization of 150 biomedical scientific articles from the BioMed Central corpus and it is found that

Proceedings of the 2010 Workshop on Biomedical Natural Language Processing, ACL 2010, pages 55–63,
Uppsala, Sweden, 15 July 2010.

WSD improves the quality of the summaries. This paper is, to our knowledge, the first to apply WSD to the summarization of biomedical documents and also demonstrates that this leads to an improvement in performance.

The next section describes related work on summarization and WSD. Section 3 introduces the UMLS resources used in the WSD and summarization systems. Section 4 describes our concept-based summarization algorithm. Section 5 presents a graph-based WSD algorithm which has been adapted to assign concepts from the UMLS. Section 6 describes the experiments carried out to evaluate the impact of WSD and discusses the results. The final section provides concluding remarks and suggests future lines of work.

2 Related work

Summarization has been an active area within NLP research since the 1950s and a variety of approaches have been proposed (Mani, 2001; Afantenos et al., 2005). Our focus is on graph-based summarization methods. Graph-based approaches typically represent the document as a graph, where the nodes represent text units (i.e. words, sentences or paragraphs), and the links represent cohesion relations or similarity measures between these units. The best-known work in the area is LexRank (Erkan and Radev, 2004). It assumes a fully connected and undirected graph, where each node corresponds to a sentence, represented by its *TF-IDF* vector, and the edges are labeled with the cosine similarity between the sentences. Mihalcea and Tarau (2004) present a similar method where the similarity among sentences is measured in terms of word overlaps.

However, methods based on term frequencies and syntactic representations do not exploit the semantic relations among the words in the text (i.e. synonymy, homonymy or co-occurrence). They cannot realize, for instance, that the phrases *myocardial infarction* and *heart attack* refer to the same concepts, or that *pneumococcal pneumonia* and *mycoplasma pneumonia* are two similar diseases that differ in the type of bacteria that causes them. This problem can be partially solved by dealing with concepts and semantic relations from domain-specific resources, rather than terms and lexical or syntactic relations. Consequently, some recent approaches have adapted existing methods to represent the document at a conceptual level. In particular, in the biomedical domain Reeve et al. (2007) adapt the lexical chaining approach (Barzilay and Elhadad, 1997) to work with UMLS concepts, using the MetaMap Transfer Tool to annotate these concepts. Yoo et al. (2007) represent a corpus of documents as a graph, where the nodes are the MeSH descriptors found in the corpus, and the edges represent hypernymy and co-occurrence relations between them. They cluster the MeSH concepts in the corpus to identify sets of documents dealing with the same topic and then generate a summary from each document cluster.

Word sense disambiguation attempts to solve lexical ambiguities by identifying the correct meaning of a word based on its context. Supervised approaches have been shown to perform better than unsupervised ones (Agirre and Edmonds, 2006) but need large amounts of manually-tagged data, which are often unavailable or impractical to create. Knowledge-based approaches are a good alternative that do not require manually-tagged data.

Graph-based methods have recently been shown to be an effective approach for knowledge-based WSD. They typically build a graph for the text in which the nodes represent all possible senses of the words and the edges represent different kinds of relations between them (e.g. lexico-semantic, co-occurrence). Some algorithm for analyzing these graphs is then applied from which a ranking of the senses of each word in the context is obtained and the highest-ranking one is chosen (Mihalcea and Tarau, 2004; Navigli and Velardi, 2005; Agirre and Soroa, 2009). These methods find globally optimal solutions and are suitable for disambiguating all words in a text.

One such method is Personalized PageRank (Agirre and Soroa, 2009) which makes use of the PageRank algorithm used by internet search engines (Brin and Page, 1998). PageRank assigns weight to each node in a graph by analyzing its structure and prefers ones that are linked to by other nodes that are highly weighted. Agirre and Soroa (2009) used WordNet as the lexical knowledge base and creates graphs using the entire WordNet hierarchy. The ambiguous words in the document are added as nodes to this graph and directed links are created from them to each of their possible meanings. These nodes are assigned weight in the graph and the PageRank algorithm is applied to distribute this information through the graph. The meaning of each word with the highest weight is chosen. We refer to this approach as ppr. It is efficient since it allows all ambiguous words in a document to be disambiguated simultaneously using the whole lexical knowledge base, but can be misled when two of the possible senses for an ambiguous word are related to each other in WordNet since the PageRank algorithm assigns weight to these senses rather than transferring it to related words. Agirre and Soroa (2009) also describe a variant of the approach, referred to as "word to word" (ppr_w2w), in which a separate graph is created for each ambiguous word. In these graphs no weight is assigned to the word being disambiguated so that all of the information used to assign weights to the possible senses of the word is obtained from the other words in the document. The ppr_w2w is more accurate but less efficient due to the number of graphs that have to be created and analyzed. Agirre and Soroa (2009) show that the Personalized PageRank approach performs well in comparison to other knowledge-based approaches to WSD and report an accuracy of around 58% on standard evaluation data sets.

3 UMLS

The Unified Medical Language System (UMLS) (Humphreys et al., 1998) is a collection of controlled vocabularies related to biomedicine and contains a wide range of information that can be used for Natural Language Processing. The UMLS comprises of three parts: the Specialist Lexicon, the Semantic Network and the Metathesaurus.

The **Metathesaurus** forms the backbone of the UMLS and is created by unifying over 100 controlled vocabularies and classification systems. It is organized around concepts, each of which represents a meaning and is assigned a Concept Unique Identifier (CUI). For example, the following CUIs are all associated with the term "cold": C0009443 'Common Cold', C0009264 'Cold Temperature' and C0234192 'Cold Sensation'.

The MRREL table in the Metathesaurus lists relations between CUIs found in the various sources that are used to form the Metathesaurus. This table lists a range of different types of relations, including CHD ("child"), PAR ("parent"), QB ("can be qualified by"), RQ ("related and possibly synonymous") and RO ("other related"). For example, the MRREL table states that C0009443 'Common Cold' and C0027442 'Nasopharynx' are connected via the RO relation.

The MRHIER table in the Metathesaurus lists the hierarchies in which each CUI appears, and presents the whole path to the top or root of each hierarchy for the CUI. For example, the MRHIER table states that C0035243 'Respiratory Tract Infections' is a parent of C0009443 'Common Cold'.

The **Semantic Network** consists of a set of categories (or semantic types) that provides a consistent categorization of the concepts in the Metathesaurus, along with a set of relationships (or semantic relations) that exist between the semantic types. For example, the CUI C0009443 'Common Cold' is classified in the semantic type 'Disease or Syndrome'.

The SRSTR table in the Semantic Network describes the structure of the network. This table lists a range of different relations between semantic types, including hierarchical relations (is_a) and non hierarchical relations (e.g. result_of, associated_with and co-occurs_with). For example, the semantic types 'Disease or Syndrome' and 'Pathologic Function' are connected via the is_a relation in this table.

4 Summarization system

The method presented in this paper consists of 4 main steps: (1) concept identification, (2) document representation, (3) concept clustering and topic recognition, and (4) sentence selection. Each step is discussed in detail in the following subsections.

4.1 Concept identification

The first stage of our process is to map the document to concepts from the UMLS Metathesaurus and semantic types from the UMLS Semantic Network.

We first run the MetaMap program over the text in the body section of the document[1] MetaMap (Aronson, 2001) identifies all the phrases that could be mapped onto a UMLS CUI, retrieves and scores all possible CUI mappings for each phrase, and returns all the candidates along with

[1]We do not make use of the disambiguation algorithm provided by MetaMap, which is invoked using the -y flag (Aronson, 2006), since our aim is to compare the effect of WSD on the performance of our summarization system rather than comparing WSD algorithms.

their score. The semantic type for each concept mapping is also returned. Table 1 shows this mapping for the phrase *tissues are often cold*. This example shows that MetaMap returns a single CUI for two words (*tissues* and *often*) but also returns three equally scored CUIs for *cold* (C0234192, C0009443 and C0009264). Section 5 describes how concepts are selected when MetaMap is unable to return a single CUI for a word.

Phrase: "Tissues"
Meta Mapping (1000)
1000 C0040300:Tissues (Body tissue)
Phrase: "are"
Phrase: "often cold"
MetaMapping (888)
694 C0332183:Often (Frequent)
861 C0234192:Cold (Cold Sensation)
MetaMapping (888)
694 C0332183:Often (Frequent)
861 C0009443:Cold (Common Cold)
MetaMapping (888)
694 C0332183:Often (Frequent)
861 C0009264:Cold (cold temperature)

Table 1: An example of MetaMap mapping for the phrase *Tissues are often cold*

UMLS concepts belonging to very general semantic types are discarded, since they have been found to be excessively broad or unrelated to the main topic of the document. These types are *Quantitative Concept*, *Qualitative Concept*, *Temporal Concept*, *Functional Concept*, *Idea or Concept*, *Intellectual Product*, *Mental Process*, *Spatial Concept* and *Language*. Therefore, the concept C0332183 'Often' in the previous example, which belongs to the semantic type *Temporal Concept*, is discarded.

4.2 Document representation

The next step is to construct a graph-based representation of the document. To this end, we first extend the disambiguated UMLS concepts with their complete hierarchy of hypernyms and merge the hierarchies of all the concepts in the same sentence to construct a graph representing it. The two upper levels of these hierarchies are removed, since they represent concepts with excessively broad meanings and may introduce noise to later processing.

Next, all the sentence graphs are merged into a single document graph. This graph is extended with more semantic relations to obtain a more complete representation of the document. Various types of information from the UMLS can be used to extend the graph. We experimented using different sets of relations and finally used the *hypernymy* and *other related* relations between concepts from the Metathesaurus, and the *associated with* relation between semantic types from the Semantic Network. Hypernyms are extracted from the `MRHIER` table, `RO` ("other related") relations are extracted from the `MRREL` table, and `associated_with` relations are extracted from the `SRSTR` table (see Section 3). Finally, each edge is assigned a weight in [0, 1]. This weight is calculated as the ratio between the relative positions in their corresponding hierarchies of the concepts linked by the edge.

Figure 1 shows an example graph for a simplified document consisting of the two sentences below. Continuous lines represent *hypernymy* relations, dashed lines represent *other related* relations and dotted lines represent *associated with* relations.

1. The goal of the trial was to assess cardiovascular mortality and morbidity for stroke, coronary heart disease and congestive heart failure, as an evidence-based guide for clinicians who treat hypertension.
2. The trial was carried out in two groups: the first group taking doxazosin, and the second group taking chlorthalidone.

4.3 Concept clustering and topic recognition

Our next step consists of clustering the UMLS concepts in the document graph using a *degree-based clustering* method (Erkan and Radev, 2004). The aim is to construct sets of concepts strongly related in meaning, based on the assumption that each of these sets represents a different topic in the document.

We assume that the document graph is an instance of a *scale-free network* (Barabasi and Albert, 1999). A scale-free network is a complex network that (among other characteristics) presents a particular type of node which are highly connected to other nodes in the network, while the remaining nodes are quite unconnected. These highest-degree nodes are often called *hubs*. This scale-free power-law distribution has been empirically observed in many large networks, including linguistic and semantic ones.

To discover these prominent or hub nodes, we compute the *salience* or prestige of each vertex

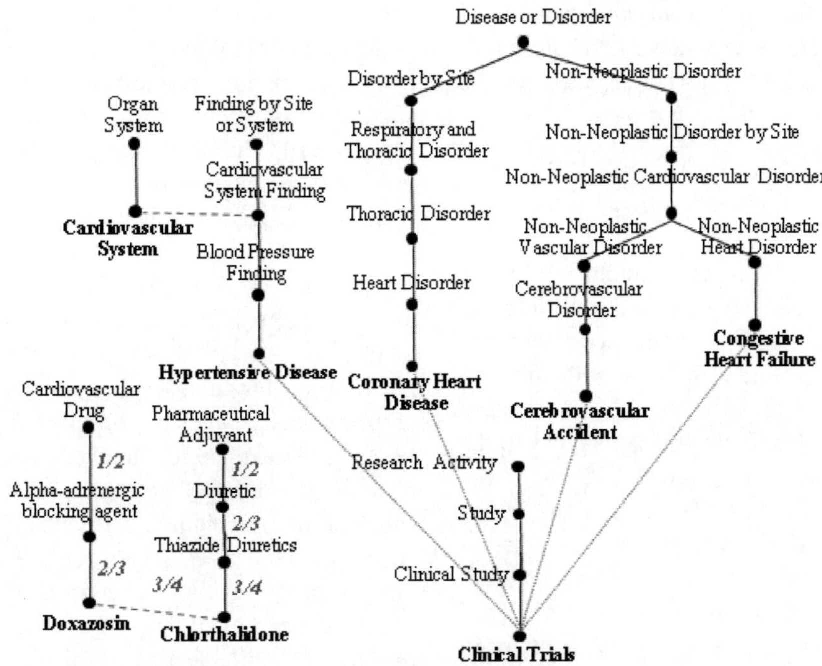

Figure 1: Example of a simplified document graph

in the graph (Yoo et al., 2007), as shown in (1). Whenever an edge from v_i to v_j exists, a vote from node *i* to node *j* is added with the strength of this vote depending on the weight of the edge. This ranks the nodes according to their structural importance in the graph.

$$salience(v_i) = \sum_{\forall e_j | \exists v_k \wedge e_j connect(v_i, v_k)} weight(e_j) \quad (1)$$

The *n* vertices with a highest salience are named *Hub Vertices*. The clustering algorithm first groups the hub vertices into *Hub Vertices Sets (HVS)*. These can be seen as set of concepts strongly related in meaning, and will represent the centroids of the clusters. To construct these HVS, the clustering algorithm first searches, iteratively and for each hub vertex, the hub vertex most connected to it, and merges them into a single HVS. Second, the algorithm checks, for every pair of HVS, if their internal connectivity is lower than the connectivity between them. If so, both HVS are merged. The remaining vertices (i.e. those not included in the HVS) are iteratively assigned to the cluster to which they are more connected. This connectivity is computed as the sum of the weights of the edges that connect the target vertex to the other vertices in the cluster.

4.4 Sentence selection

The last step of the summarization process consists of computing the similarity between all sentences in the document and each of the clusters, and selecting the sentences for the summary based on these similarities. To compute the similarity between a sentence graph and a cluster, we use a non-democratic vote mechanism (Yoo et al., 2007), so that each vertex of a sentence assigns a vote to a cluster if the vertex belongs to its HVS, half a vote if the vertex belongs to it but not to its HVS, and no votes otherwise. Finally, the similarity between the sentence and the cluster is computed as the sum of the votes assigned by all the vertices in the sentence to the cluster, as expressed in (2).

$$similarity(C_i, S_j) = \sum_{v_k | v_k \in S_j} w_{k,j} \quad (2)$$

$$\text{where} \begin{cases} w_{k,j}=0 \; if \; v_k \notin C_i \\ w_{k,j}=1 \; if \; v_k \in HVS(C_i) \\ w_{k,j}=0.5 \; if \; v_k \notin HVS(C_i) \end{cases}$$

Finally, we select the sentences for the summary based on the similarity between them and the clusters as defined above. In previous work (blind reference), we experimented with different heuristics for sentence selection. In this paper, we just present the one that reported the best results. For each sentence, we compute a single score, as the sum of its similarity to each cluster adjusted to the cluster's size (expression 3). Then, the N sentences with higher scores are selected for the summary.

$$Score(S_j) = \sum_{C_i} \frac{similarity(C_i, S_j)}{|C_i|} \quad (3)$$

In addition to semantic-graph similarity (*SemGr*) we have also tested two further features for computing the salience of sentences: sentence location (*Location*) and similarity with the title section (*Title*). The sentence location feature assigns higher scores to the sentences close to the beginning and the end of the document, while the similarity with the title feature assigns higher scores as the proportion of common concepts between the title and the target sentence is increased. Despite their simplicity, these are well accepted summarization heuristics that are commonly used (Bawakid and Oussalah, 2008; Bossard et al., 2008).

The final selection of the sentences for the summary is based on the weighted sum of these feature values, as stated in (4). The values for the parameters λ, θ and χ have been empirically set to 0.8, 0.1, and 0.1 respectively.

$$Score(S_j) = \lambda \times SemGr(S_j) + \\ \theta \times Location(S_j) + \chi \times Title(S_j) \quad (4)$$

5 WSD for concept identification

Since our summarization system is based on the UMLS it is important to be able to accurately map the documents onto CUIs. The example in Section 4.1 shows that MetaMap does not always select a single CUI and it is therefore necessary to have some method for choosing between the ones that are returned. Summarization systems typically take the first mapping as returned by MetaMap, and no attempt is made to solve this ambiguity (Plaza et al., 2008). This paper reports an alternative approach that uses a WSD algorithm that makes use of the entire UMLS Metathesaurus.

The Personalized PageRank algorithm (see Section 2) was adapted to use the UMLS Metathesaurus and used to select a CUI from the MetaMap output[2]. The UMLS is converted into a graph in which the CUIs are the nodes and the edges are derived from the MRREL table. All possible relations in this table are included. The output from MetaMap is used to provide the list of possible CUIs for each term in a document and these are passed to the disambiguation algorithm. We use both the standard (ppr) and "word to word" (ppr_w2w) variants of the Personalized PageRank approach.

It is difficult to evaluate how well the Personalized PageRank approach performs when used in this way due to a lack of suitable data. The NLM-WSD corpus (Weeber et al., 2001) contains manually labeled examples of ambiguous terms in biomedical text but only provides examples for 50 terms that were specifically chosen because of their ambiguity. To evaluate an approach such as Personalized PageRank we require documents in which the sense of every ambiguous word has been identified. Unfortunately no such resource is available and creating one would be prohibitively expensive. However, our main interest is in whether WSD can be used to improve the summaries generated by our system rather than its own performance and, consequently, decided to evaluate the WSD by comparing the output of the summarization system with and without WSD.

6 Experiments

6.1 Setup

The ROUGE metrics (Lin, 2004) are used to evaluate the system. ROUGE compares automatically generated summaries (called *peers*) against human-created summaries (called *models*), and calculates a set of measures to estimate the content quality of the summaries. Results are reported for the ROUGE-1 (**R-1**), ROUGE-2 (**R-2**), ROUGE-SU4 (**R-SU**) and ROUGE-W (**R-W**) metrics. ROUGE-N (e.g. ROUGE-1 and ROUGE-2) evaluates n-gram co-occurrences among the peer and models summaries, where N stands for the length of the n-grams. ROUGE-SU4 allows bi-gram to have intervening word gaps no longer than four words. Finally, ROUGE-W computes the union of the longest common subsequences between the candidate and the reference summaries taking into account the presence of consecutive matches.

To the authors' knowledge, no specific corpus for biomedical summarization exists. To evaluate our approach we use a collection of 150 documents randomly selected from the BioMed Cen-

[2]We use a publicly available implementation of the Personalized Page Rank algorithm (http://ixa2.si.ehu.es/ukb/) for the experiments described here.

tral corpus[3] for text mining research. This collection is large enough to ensure significant results in the ROUGE evaluation (Lin, 2004) and allows us to work with the `ppr_w2w` disambiguation software, which is quite time consuming. We generate automatic summaries by selecting sentences until the summary reaches a length of the 30% over the original document size. The abstract of the papers (i.e. the authors' summaries) are removed from the documents and used as model summaries.

A separate development set was used to determine the optimal values for the parameters involved in the algorithm. This set consists of 10 documents from the BioMed Central corpus. The model summaries for these documents were manually created by medical students by selecting between 20-30% of the sentences within the paper. The parameters to be estimated include the percentage of vertices considered as hub vertices by the clustering method (see Section 4.3) and the combination of summarization features used to sentence selection (see Section 4.4). As a result, the percentage of hub vertices was set to 15%, and no additional summarization features (apart from the semantic-graph similarity) were used.

Two baselines were also implemented. The first, *lead baseline*, generate summaries by selecting the first *n* sentences from each document. The second, *random baseline*, randomly selects *n* sentences from the document. The *n* parameter is based on the desired compression rate (i.e. 30% of the document size).

6.2 Results

Various summarizers were created and evaluated. First, we generated summaries using our method without performing word sense disambiguation (SemGr), but selecting the first CUI returned by MetaMap. Second, we repeated these experiments using the Personalized Page Rank disambiguation algorithm (`ppr`) to disambiguate the CUIs returned by MetaMap (SemGr + `ppr`). Finally, we use the "word to word" variant of the Personalized Page Rank algorithm (`ppr_w2w`) to perform the disambiguation (SemGr + `ppr_w2w`).

Table 2 shows ROUGE scores for the different configurations of our system together with the two baselines. All configurations significantly outperform both baselines (Wilcoxon Signed Ranks Test, $p < 0.01$).

[3] http://www.biomedcentral.com/info/about/datamining/

Summarizer	R-1	R-2	R-W	R-SU
random	.5089	.1879	.1473	.2349
lead	.6483	.2566	.1621	.2646
SemGr	.7504	.3283	.1915	.3117
SemGr+`ppr`	**.7737**	**.3419**	.1937	.3178
SemGr+`ppr_w2w`	**.7804**	**.3530**	**.1966**	**.3262**

Table 2: ROUGE scores for two baselines and SemGr (with and without WSD). Significant differences among the three versions of SemGr are indicated in bold font.

The use of WSD improves the average ROUGE score for the summarizer. The "standard" (i.e. `ppr`) version of the WSD algorithm significantly improves ROUGE-1 and ROUGE-2 metrics (Wilcoxon Signed Ranks Test, $p < 0.01$), compared with no WSD (i.e. SemGr). The "word to word" variant (`ppr_w2w`) significantly improves all ROUGE metrics. Performance using the "word to word" variant is also higher than standard `ppr` in all ROUGE scores.

These results demonstrate that employing a state of the art WSD algorithm that has been adapted to use the UMLS Metathesaurus improves the quality of the summaries generated by a summarization system. To our knowledge this is the first result to demonstrate that WSD can improve summarization systems. However, this improvement is less than expected and this is probably due to errors made by the WSD system. The Personalized PageRank algorithms (`ppr` and `ppr_w2w`) have been reported to correctly disambiguate around 58% of words in general text (see Section 2) and, although we were unable to quantify their performance when adapted for the biomedical domain (see Section 5), it is highly likely that they will still make errors. However, the WSD performance they do achieve is good enough to improve the summarization process.

6.3 Analysis

The results presented above demonstrate that using WSD improves the performance of our summarizer. The reason seems to be that, since the accuracy in the concept identification step increases, the document graph built in the following steps is a better approximation of the structure of the document, both in terms of concepts and relations. As a result, the clustering method succeeds in finding the topics covered in the document, and the information in the sentences selected for the summary is closer to that presented in the model summaries.

We have observed that the clustering method usually produces one big cluster along with a variable number of small clusters. As a consequence, though the heuristic for sentence selection was designed to select sentences from all the clusters in the document, the fact is that most of the sentences are extracted from this single large cluster. This allows our system to identify sentences that cover the main topic of the document, while it occasionally fails to extract other "satellite" information.

We have also observed that the ROUGE scores differ considerably from one document to others. To understand the reasons of these differences we examined the two documents with the highest and lowest ROUGE scores respectively. The best case is one of the largest document in the corpus, while the worst case is one of the shortest (6 versus 3 pages). This was expected, since according to our hypothesis that the document graph is an instance of a scale-free network (see Section 4.3), the summarization algorithm works better with larger documents. Both documents also differ in their underlying subject matter. The best case concerns the reactions of some kind of proteins over the brain synaptic membranes; while the worst case regards the use of pattern matching for database searching. We have verified that UMLS covers the vocabulary contained in the first document better than in the second one. We have also observed that the use in the abstract of synonyms of terms presented in the document body is quite frequent. In particular the worst case document uses different terms in the abstract and the body, for example "pattern matching" and "string searching". Since the ROUGE metrics rely on evaluating summaries based on the number of strings they have in common with the model summaries the system's output is unreasonably penalised.

Another problem is related to the use of acronyms and abbreviations. Most papers in the corpus do not include an *Abbreviations* section but define them *ad hoc* in the document body. These contracted forms are usually non-standard and do not exist in the UMLS Metathesaurus. This seriously affects the performance of both the disambiguation and the summarization algorithms, especially considering that it has been observed that the terms (or phrases) represented in an abbreviated form frequently correspond to central concepts in the document. For example, in a paper from the corpus that presents an analysis tool for simple sequence repeat tracts in DNA, only the first occurrence of 'simple sequence repeat' is presented in its expanded form. In the remaining of the document, this phrase is named by its acronym 'SSR'. The same occurs in a paper that investigates the developmental expression of survivin during embryonic submandibular salivary gland development, where 'embryonic submandibular gland' is always referred as 'SMG'.

7 Conclusion and future work

In this paper we propose a graph-based approach to biomedical summarization. Our algorithm represents the document as a semantic graph, where the nodes are concepts from the UMLS Metathesaurus and the links are different kinds of semantic relations between them. This produces a richer representation than the one provided by traditional models based on terms.

This approach relies on accurate mapping of the document being summarized into the concepts in the UMLS Metathesaurus. Three methods for doing this were compared and evaluated. The first was to select the first mapping generated by MetaMap while the other two used a state of the art WSD algorithm. This WSD algorithm was adapted for the biomedical domain by using the UMLS Metathesaurus as a knowledge based and MetaMap as a pre-processor to identify the possible CUIs for each term. Results show that the system performs better when WSD is used.

In future work we plan to make use of the different types of information within the UMLS to create different configurations of the Personalized PageRank WSD algorithm and explore their effect on the summarization system (i.e. considering different UMLS relations and assigning different weights to different relations). It would also be interesting to test the system with other disambiguation algorithms and use a state of the art algorithm for identifying and expanding acronyms and abbreviations.

Acknowledgments

This research is funded by the Spanish Government through the FPU program and the projects TIN2009-14659-C03-01 and TSI 020312-2009-44. Mark Stevenson acknowledges the support of the Engineering and Physical Sciences Research Council (grant EP/D069548/1).

References

S.D. Afantenos, V. Karkaletsis, and P. Stamatopoulos. 2005. Summarization from medical documents: a survey. *Artificial Intelligence in Medicine*, 33(2):157–177.

E. Agirre and P. Edmonds, editors, 2006. *Word Sense Disambiguation: Algorithms and Applications*. Springer.

E. Agirre and A. Soroa. 2009. Personalizing PageRank for Word Sense Disambiguation. In *Proceedings of EACL-09*, pages 33–41, Athens, Greece.

A. Aronson. 2001. Effective mapping of biomedical text to the UMLS Metathesaurus: the MetaMap program. In *Proceedings of the AMIA Symposium*, pages 17–21.

A. Aronson. 2006. MetaMap: Mapping text to the UMLS Metathesaurus. Technical report, U.S. National Library of Medicine.

A.L. Barabasi and R. Albert. 1999. Emergence of scaling in random networks. *Science*, 268:509–512.

R. Barzilay and M. Elhadad. 1997. Using lexical chains for text summarization. In *Proceedings of the ACL Workshop on Intelligent Scalable Text Summarization*, pages 10–17.

A. Bawakid and M. Oussalah. 2008. A semantic summarization system: University of Birmingham at TAC 2008. In *Proceedings of the First Text Analysis Conference (TAC 2008)*.

A. Bossard, M. Gnreux, and T. Poibeau. 2008. Description of the LIPN systems at TAC 2008: summarizing information and opinions. In *Proceedings of the First Text Analysis Conference (TAC 2008)*.

S. Brin and L. Page. 1998. The anatomy of a large-scale hypertextual web search engine. *Computer Networks and ISDN Systems*, 30:1–7.

G. Erkan and D. R. Radev. 2004. LexRank: Graph-based lexical centrality as salience in text summarization. *Journal of Artificial Intelligence Research (JAIR)*, 22:457–479.

M. Fiszman, T. C. Rindflesch, and H. Kilicoglu. 2004. Abstraction summarization for managing the biomedical research literature. In *Proceedings of the HLT-NAACL Workshop on Computational Lexical Semantics*, pages 76–83.

L. Humphreys, D. Lindberg, H. Schoolman, and G. Barnett. 1998. The Unified Medical Language System: An informatics research collaboration. *Journal of the American Medical Informatics Association*, 1(5):1–11.

L. Hunter and K. B. Cohen. 2006. Biomedical Language Processing: Perspective Whats Beyond PubMed? *Mol Cell.*, 21(5):589–594.

C.-Y. Lin. 2004. Rouge: A package for automatic evaluation of summaries. In *Proceedings of the ACL-04 Workshop: Text Summarization Branches Out.*, pages 74–81, Barcelona, Spain.

I. Mani. 2001. *Automatic summarization*. Jonh Benjamins Publishing Company.

R. Mihalcea and P. Tarau. 2004. TextRank - Bringing order into text. In *Proceedings of the Conference EMNLP 2004*, pages 404–411.

R. Navigli and P. Velardi. 2005. Structural semantic interconnections: A knowledge-based approach to word sense disambiguation. *IEEE Trans. Pattern Anal. Mach. Intell.*, 27(7):1075–1086.

S. Nelson, T. Powell, and B. Humphreys. 2002. The Unified Medical Language System (UMLS) Project. In Allen Kent and Carolyn M. Hall, editors, *Encyclopedia of Library and Information Science*. Marcel Dekker, Inc.

L. Plaza, A. Díaz, and P. Gervás. 2008. Concept-graph based biomedical automatic summarization using ontologies. In *TextGraphs '08: Proceedings of the 3rd Textgraphs Workshop on Graph-Based Algorithms for Natural Language Processing*, pages 53–56.

L.H. Reeve, H. Han, and A.D. Brooks. 2007. The use of domain-specific concepts in biomedical text summarization. *Information Processing and Management*, 43:1765–1776.

M. Weeber, J. Mork, and A. Aronson. 2001. Developing a Test Collection for Biomedical Word Sense Disambiguation. In *Proceedings of AMIA Symposium*, pages 746–50, Washington, DC.

I. Yoo, X. Hu, and I-Y. Song. 2007. A coherent graph-based semantic clustering and summarization approach for biomedical literature and a new summarization evaluation method. *BMC Bioinformatics*, 8(9).

Cancer Stage Prediction Based on Patient Online Discourse

Mukund Jha
Computer Science
Columbia University
New York, NY 10027
mj2472@columbia.edu

Noémie Elhadad
Biomedical Informatics
Columbia University
New York, NY 10032
noemie@dbmi.columbia.edu

Abstract

Forums and mailing lists dedicated to particular diseases are increasingly popular online. Automatically inferring the health status of a patient can be useful for both forum users and health researchers who study patients' online behaviors. In this paper, we focus on breast cancer forums and present a method to predict the stage of patients' cancers from their online discourse. We show that what the patients talk about (content-based features) and whom they interact with (social network-based features) provide complementary cues to predicting cancer stage and can be leveraged for better prediction. Our methods are extendable and can be applied to other tasks of acquiring contextual information about online health forum participants.

1 Introduction

In this paper we investigate an automated method of inferring the stage of a patient's breast cancer from discourse in an online forum. Such information can prove invaluable both for forum members, by enriching their use of this rapidly developing and increasingly popular medium, and for health researchers, by providing them with tools to quantify and better understand patient populations and how they behave online.

Patients with chronic diseases like diabetes or life-threatening conditions like breast cancer get a wealth of information from medical professionals about their diagnoses, test results, and treatment options, but such information is not always satisfactory or sufficient for patients. Much of that is essential to their everyday lives and the management of their condition escapes the clinical realm. Furthermore, patients feel informed and empowered by exchanging experiences and emotional support with others in the same circumstances. Thus, it is not surprising that patient communities have flourished on the Web over the past decade, through active disease-specific discussion forums and mailing lists.

For health professionals, this new medium presents exciting research avenues related to theories of psycho-social support and how patients manage their conditions. Qualitative analyses of forums and mailing list posts show that breast cancer patients and survivors provide and seek support to and from their peers and that support, while also emotional, is largely informational in nature (Civan and Pratt, 2007; Meier et al., 2007). Emotional support may include words of encouragement and prayers. Examples of informational support are providing personal experiences with a treatment, discussing new research, explaining a pathology report to a peer, as well as exchanging information pertinent to patients' daily lives, such as whether to shave one's head once chemotherapy starts.

Given the kinds of benefits that patients and survivors seek and provide in online forums, it seems likely that they would be inclined to gravitate toward others whose circumstances most closely resemble their own, beyond sharing the general diagnosis of breast cancer. In fact, focus groups and surveys conducted with breast cancer patients identified and emphasized the need for online cancer forum participants to identify other patients of a particular age, stage of illness, or having opted for similar treatment (Rozmovits and Ziebland, 2004; van Uden-Kraan et al., 2008).

The stage of a patient's cancer, in particular, can be a crucial proxy for finding those whose experiences are likely similar and relevant to one's own. For breast cancer, there are five high-level standard stages (0 to IV). While they do not give the whole picture about a particular cancer (the stages

Proceedings of the 2010 Workshop on Biomedical Natural Language Processing, ACL 2010, pages 64–71,
Uppsala, Sweden, 15 July 2010.

themselves can be described with finer granularity and they do no not encompass additional information like hormonal sensitivity), physicians have traditionally relied on them for prognosis and determining treatment options. For patients and survivors, they are a useful way to communicate to their peers their health status, as evidenced by the members' signatures on forums and mailing lists (Meier et al., 2007).

Although many forums provide pre-set profile fields for users to populate with important background information, such as the stage of their cancer (e.g., the popular forum on `breastcancer.org`), in practice, only a fraction of members have a complete profile. Thus, an automated way of inferring member profile information via the social network created by a forum's users would help fill in the blanks.

Beyond identifying other patients in a forum in similar circumstances, such a tool can have numerous practical benefits for both forum users and health researchers who study patients' online behavior. When a patient searches for a particular piece of information in a forum, incorporating contextual information about the user into the search mechanism can improve search results. For example, a search tool can rank higher the posts that were authored by patients with the same stage. For health researchers, questions which bring a better understanding of forum usage (i.e., "are patients with stage IV cancer more or less active in a forum than patients with early stage cancer") can be answered accurately only if all members of the forums are taken into account, not just the ones who filled out their member profiles. Furthermore, in the context of health communication, the more information is available about an individual, the more effective the message can be, from generic to personalized to targetted to tailored (Kreuter et al., 2000). Our research contributes an automated method to acquiring contextual information about forum participants. We focus on cancer stage as an exmple of context information.

Our research question is whether it is possible to predict the stage of individuals' cancer based on their online discourse. By discourse we mean both the information she conveys and whom she talks to in a forum. Following ethical guidelines in processing of patient data online, we focus on a popular breast cancer forum with a large number of participants (Eysenbach and Till, 2001). We show that the content of members' posts and the stage of their interlocutors can provide complementary clues to identifying cancer stages.

2 Related Work

Researchers have begun to explore the possibility of diagnosing patients based on their speech productions. Content analysis methods, which rely on patient speech transcripts or texts authored by patients, have been leveraged for understanding cancer coping mechanisms (Graves et al., 2005; Bantum and Owen, 2009), psychiatric diagnoses (Oxman et al., 1988; Elvevaag et al., 2010), and the analysis of suicide notes (Pestian et al., 2008). In all cases, results, while not fully accurate, are promising and show that patient-generated content is a valuable clue to diagnosis in an automated framework.

Our work departs from these experiments in that we do not attempt to predict the psychological state of a patient, but rather the status of a clinical condition. Staging breast cancer provides a way to summarize the status of the cancer based on clinical characteristics (the size of the tumor, whether the cancer is invasive or not, whether cancer cells are present in the lymph nodes, and whether the cancer has spread beyond the breast). There are five high-level stages for breast cancer. Stage 0 describes a non-invasive cancer. Stage I represents early stage of an invasive cancer, where the tumor size is less than 2 centimeters and no lymph nodes are involved (that is, the cancer has not spread outside of the breast). Stages II and III describe a cancer with larger tumor size and/or the cancer has spread outside of the breast. Stage IV describes a cancer that have metastasized to distant parts of the body, such as lungs and bones.

In our work, we analyze naturally occurring content, generated by patients talking to each other online. As such, our sample population is much larger than in earlier works (typically less than 100 subjects). Like the researchers who focus on content analysis, we rely on the content generated by patients, but we also hypothesize that whom the patients interact with can help the prediction of cancer stage.

In particular, we build a social network based on patients' interactions to boost text-based predictions. Graph-based methods are becoming increasingly popular in the NLP community, and similar approaches have been employed and shown to perform well in other areas like question answering (Jurczyk, 2007) (Harabagiu et al., 2006), word-sense disambiguation (Niu et al., 2005), and textual entailment (Haghighi, 2005).

3 Methods

Our methods to predict cancer stage operate in a supervised framework. We cast the task of stage prediction as a 4-way classification (Stage I to IV). We hypothesize that the discourse of patients online, as defined by the content of their posts in a forum, can be leveraged to predict cancer stage. Furthermore, we hypothesize that the social network derived by whom patients interact with can provide an additional clue for stage detection.

We experimented with three methods of predicting cancer stage:

Text-based stage prediction A classifier is trained given the post history of a patient.

Network-based stage prediction A social network representing the interactions among forum members is built, and a label propagation algorithm is applied to infer the stage of individual patients.

Combined prediction A classifier which combines text-based and network-based features.

Next we describe each method in detail, along with our dataset and our experimental setup.

3.1 Data Collection and Preprocessing

We collected posts from the publicly available discussion board from `breastcancer.org`. It is a popular forum, with more than 60,000 registered members, and more than 50,000 threads discussed in 60 subforums. To collect our dataset, we crawled the content of the most popular subforums.[1]

Collected posts were translated from HTML into an XML format, keeping track of author id, thread id, position of the post in the thread, body of the post, and signature of the author (which is kept separated from the body of the post). The content of the posts was tokenized, lower-cased and stemmed. Images, URLs, and stop words were removed.

Nb. of threads	26,160
Nb. of posts	524,247
Nb. of threads with $<$ 20 posts	22,334
Nb. of users with profile Stage I	2,226
Nb. of users with profile Stage II	2,406
Nb. of users with profile Stage III	1,031
Nb. of users with profile Stage IV	749
Total Nb. of users with profile	6,412
Nb. of active users profiled Stage I	1,317
Nb. of active users profiled Stage II	1,400
Nb. of active users profiled Stage III	580
Nb. of active users profiled Stage IV	448
Total Nb. of active users with profile	3,745

Table 1: General statistics of the dataset.

To post in `breastcancer.org`, users must register. They have the option to enter a profile with pre-set fields related to their breast cancer diagnosis; in particular cancer stage between stage I and IV. We collected the list of members who entered their stage information, thereby providing us with an annotated set of patients with their corresponding cancer stage. Table 1 shows various statistics for our dataset. Active users are defined as members who have posted more than 50 words overall in the forums. Note the low number of user with profile information (approximately 10% of the overall number of registered participants in the forum).

3.2 Text-Based Stage Prediction

We trained a text-based classifier relying on the full post history of each patient. The full post history was concatenated. Signature information, which is derived automatically from the patient's profile (and thus contains stage information) was removed from the posts. The classifier relied on unigrams and bigrams only. Table 2 shows statistics about post history length, measured as number of words authored by a forum member.

3.3 Network-Based Stage Prediction

We hypothesize that patients tend to interact in a forum with patients with similar stage. To test this

[1]There were 17 such subforums: "Just Diagnosed," "Help Me Get Through Treatment," "Surgery - Before, During, and After," "Chemotherapy - Before, During and After," "Radiation Therapy - Before, During and After," "Hormonal Therapy - Before, During and After," "Alternative, Complementary and Holistic Treatment," "Stage I and II Breast Cancer," "Just Diagnosed with a Recurrence or Metastasis," "Stage III Breast Cancer," "Stage IV Breast Cancer Survivors," "HER2/neu Positive Breast Cancer," "Deperession, Anxiety and Post Traumatic Stress Disorder," "Fitness and Getting Back in Shape," "Healthy Recipes for Everyday Living," "Recommend Your Resources," "Clinical Trials, Research, News, and Study Results."

Stages	Min	Max	Average	Median
I	4	609,608	8,429	3,123
II	2	353,731	8,142	3,112
III	8	211,655	9,297	3,189
IV	10	893,326	17,083	326

Table 2: Statistics about number of words in post history.

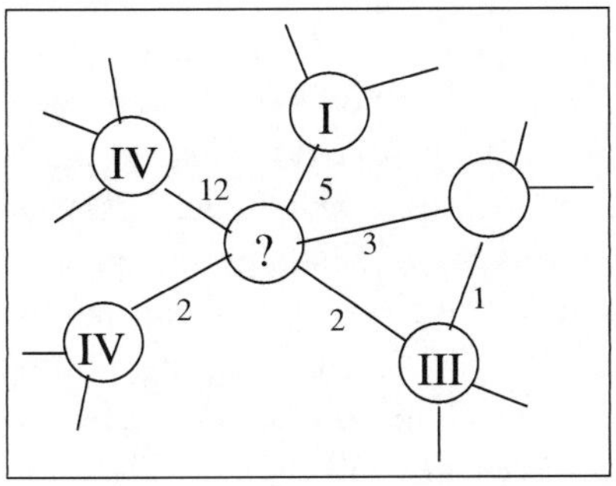

Figure 1: Nodes in the social network of forum member interaction.

hypothesis, we represent the interactions of the patients as a social network. The nodes in the network represent patients, and an edge is present between two nodes if the patients interact with each other, that is they are part of the same threads often. Weights on edges represent the degree of interaction. Higher weight on an edge between two forum members indicates they interact more often. More precisely, we build an undirected, weighted network, where the nodes representing training instances are labeled with their provided stage information and their labels are fixed. Figure 1 shows an example of node and its immediate neighbors in the network. Of his five neighbors, four represent training instances and have a fixed stage, and one represents a user with an unknown stage.

A label propagation algorithm is applied to the network, so that every node in the network is assigned a stage between I and IV (Raghavan et al., 2007). Given a node and its immediate neighbors, it looks for the most frequent labels, taking into account the edge weights. In our example, the propagated label for the central node will be stage IV. This label, in turn, will be used to assign a label to the other nodes. When building the social network of interactions, we experimented with the following parameters.

Nodes in the network. We experimented with including all the forum members who participated in a conversation thread. Thus, it includes all the members, even the ones without a known cancer stage. This resulted in a network of 15,035 forum participants. This way, the network covers more interactions among more users, but is very sparse in its initial labeling (only the training instances in the dataset of active members with a known label are labeled). The label propagation algorithm assigns labels to all the nodes, but we test its accuracy only on the test instances. We also experimented with including only the patients in the training and testing sets, thereby reducing the size of the network but also decreasing the sparsity of the labeling. This resulted in a network of 3,305 nodes.[2]

Drawing edges in the network. An edge between two users indicate they are frequently interacting. One crude way is to draw an edge between every user participating in the same thread, this however does not provide an accurate picture and hence does not yield good results. In our approach we draw an edge in two steps. First, since threads are often long and can span over multiple topics, we only draw an edge if the two individuals' posts are within five posts of each other in the thread. Second, we then look for any direct references made by a user to another user in their post. In forum threads, users usually make a direct reference by either by explicitly referring to each other using their real name or internet aliases or by quoting each other, i.e., repeating or stating what the other user has mentioned in her post. For example in *"Hey Dana, I went through the same thing the first time I went to my doctor..."*, the author of the post is referring to another user with name *'Dana'*. We rely on such explicit references to build accurate graph.[3] To find direct explicit references, we search in every post of a thread for any mention of names (real or aliases) of users participating in the thread and if one is found we draw an edge between them.

We observed that users refer to each other very

[2]This number of nodes is less than the numbers of overall active members in our gold standard because some active members have either posted in threads with only one post or with more than 20 posts.

[3]An alternative approach is to identify quotes in posts. In our particular dataset, quotes did not occur often, and thus were ignored when assessing the degree of interaction between two forum members.

frequently using their real names instead of internet names (which are long and often arbitrary). These are often hard to detect because no data is present which link users' forum aliases to their real name. We use following approach to extract real names of the users.

Extracting real names. For every user, we extract the last ten words (signature) from every post posted by the user and concatenate them after removing all stop words and other common signature terms (like thanks, all the best, love, good luck etc.) using a pre-compiled list. We then mine for the most frequent name occurring in the concatenated text using standard list of names and extracting capitalized words. We also experimented with using Named Entity Recognizers, but our simple rule based name extractor gave us better results with higher precision. Finally, we map the extracted real name with the user's alias and utilize them to find direct references between posts.

Weights Computation. The weight of an edge between two nodes represents the degree of interaction between two corresponding users (the more often they communicate, the higher the weight). Since the label propagation algorithm takes into account the weighted frequency of neighboring nodes, these weights are crucial. We compute the weights in following manner: for each pair of users with an existing edge (as determined above), we iterate through their posts in common threads, and add the cosine similarity score between the two posts to the weight of the edge. For edges made through direct references we add the highest cosine similarity score between any two pair of posts in that particular thread. This way we weigh higher the edges made through direct reference as we are more confident about them.

The full network of all users (15,035 nodes) had 480,051 edges, and the restricted network of dataset users (3,305 nodes) had 28,152 edges.

3.4 Combining Text-Based and Network-Based Predictions

To test the hypothesis that text-based and network-based predictions model different aspects of patients and thus provide complementary cues to stage prediction, we trained a classifier which incorporates text-based and network-based features.

The combined classifier contained the following features: text-based predicted label, confidence score of the text-based prediction, network-based predicted label, percentage of immediate neighbors in the network with a stage I label, stage II, III and IV labels (neighbors in the network with no labels do not contribute to the counts). For instance, the central node in Figure 1 is assigned the feature values 1/4, 0, 1/4 and 1/2 for the ratio of stage I, II, III and IV neighbors.

3.5 Experimental Setup

Our dataset for the three models consisted of the 3,745 active members. For all the models, we follow a five-fold stratified cross validation scheme. The text-based classification was carried out with BoosTexter (Schapire and Singer, 2000), trained with 800 rounds of boosting. The label propagation on the social network was carried out in R.[4] The final decision-tree classification was carried out in Weka, relying on an SVM classifier with default parameters (Hall et al., 2009).

4 Results

Table 3 shows the results of the text-based prediction, the network-based prediction and the combined prediction for each stage measured by Precision, Recall and F-measure. For comparison, we report on the results of a baseline text-based prediction. The baseline prediction assigns a stage based on the explicit mention of stage in the post history of a patient. In practice, it is a rule-based prediction with matching against the pattern "`stage [IV|four|4]`" for stage IV prediction, and similarly for other stages. The text-based prediction yields better results than the baseline, with a marked improvement for each stage.

The network-based prediction performs only slightly worse than the text-based predictions. The hypothesis that whom the patient interacts with in the forums helps predict stage holds. To verify this point further, we computed for each stage the average ratio of neighbors per stage based on the social network of interactions, as shown in Figure 2. For instance, stage IV patients interact mostly with their peers (49% of their posts are shared with other stage IV users), and to some extent with other patients (18% of their posts with stage I patients, 20% with stage II patients, and 13% with stage III patients). Except for stage III patients, all other patients are mostly interacting with similarly staged patients.

[4] `www.r-project.org`

Baseline				Text Based			
Stage	Precision	Recall	F	Stage	Precision	Recall	F
I	76.2	26.4	39.3	I	54.9	63.9	59.1
II	79.4	18.7	30.3	II	51.6	55.0	53.2
III	76.6	35.0	48.0	III	52.7	30.3	38.5
IV	76.4	50.7	60.9	IV	82.5	71.2	76.4
Network Based				**Combined**			
Stage	Precision	Recall	F	Stage	Precision	Recall	F
I	50.4	56.7	53.4	I	57.1	65.4	61.0
II	49.6	49.1	49.3	II	56.6	53.5	55.0
III	65.7	27.7	39.0	III	56.1	48.3	51.9
IV	59.3	83.7	69.4	IV	84.7	81.3	83.0

Table 3: Stage prediction results (Precision, Recall, and F-measure).

When combining the text-based and the network-based predictions in an overall classifier the prediction yields the best results. These results confirm the potential in combining the two facets of patient discourse, content and social interaction.

The results presented in the table correspond to a network built with the full set of users, including those without any profile information. When restricting the network on the patients with stage labels only, we obtained similar results (F-measures of 56% for stage I, 52% for stage II, 43% for stage III, and 79% for stage IV). This shows that it is worth modeling the full set of interactions and the full network structure, even when a large number of nodes have missing labels.

Finally, we also experimented with building networks with no weights or with weights without the 5-post-apart restriction. In both cases, the results of the network-based and combined predictions are lower than those presented in Table 3. We interpret this fact as a confirmation that our edge weighting strategy models to a promising extent the degree of interaction among patients.

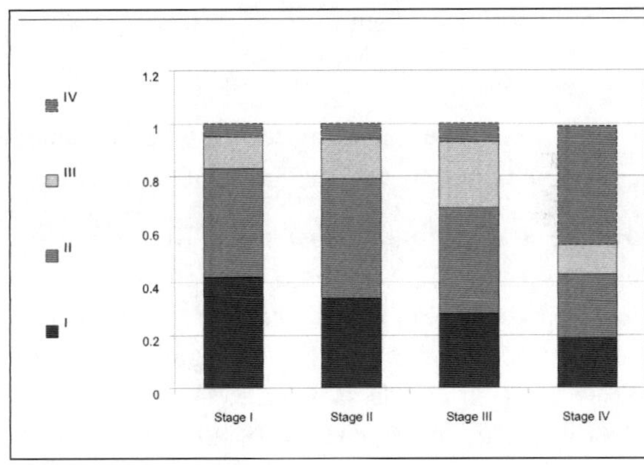

Figure 2: Distribution of stage-wise interactions.

5 Discussion

Text-based prediction. Results confirm that cancer stage can be predicted by a patient's online discourse. When examining the unigrams and bigrams picked up by the classifier as predictive of stage, we can get a sense of the frequent topics of discussion of patients. For instance, the phrases "tumor mm" (referring to tumor size in millimeters) and "breast radiation" were highly predictive of stage I patients. The words "hat" and "hair" were highly predictive of stages II and III, while stage IV patients were predicted by the presence of the phrases "bone met." (which stands for bone metastasis), "met lung" "liver," and "lymphedema" (which is a side effect of cancer treatment linked to the removal of lymph nodes and tumor).

Figure 3 shows the overall accuracy of the text-based classifier, when tested against the amount of text available for the classification. As expected, the longer the post history, the more accurate the classification.

Representing degree of interaction among patients. In our experiments, we observed that the weigthing scheme of edges had a strong impact on the overall accuracy of stage prediction. The more interaction was modeled (through distance in thread and identification of explicit references), the better the results. This confirms the hypothesis that dialogue is helpful in predicting cancer stage, and emphasizes the need for accurate techniques

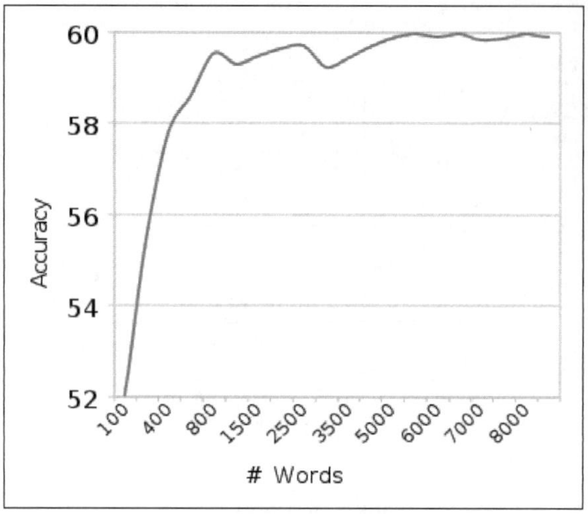

Figure 3: Overall text-based prediction accuracy against post history length.

to model interaction among forum participants in a social network.

Discourse of Stage IV patients. Both the text-based and the network-based predictions provide higher precision and recall for the stage IV patients. This is emphasized by Figure 2, where we see that, in our dataset, stage IV patients talk mostly to each other. These results suggest that stage IV patients have particular discourse, which separates them from other patients. This presents interesting avenues for future investigation.

6 Future Work and Conclusion

In this paper, we investigated breast cancer stage prediction based on the online discourse of patients participating in a breast cancer-specific forum. We show that relying on lexical features derived from the content of the posts of a patient provides promising classification results. Furthermore, even a simple social network representing patient interactions on a forum, yields predictions with comparable results. Combining the two approaches boosts results, as content and interaction seem to model complementary aspects of patient discourse.

Our experiments show that stage IV patients appear to exhibit specific textual and social patterns in forums. This point can prove useful to health researchers who want to quantify patient behaviors online.

The strategy of combining two facets of discourse (content and interactions) introduces several interesting research questions. In the future, we plan to investigate some of them. In a first step, we plan to better model the interactions of patients online. For instance, we would like to analyze the content of the posts to determine further if two patients are in direct communication, and the domain of their exchange (e.g., clinical vs. day-to-day vs. emotional). As we have observed that the way edges in the network are weighted has an impact on overall performance, we could then investigate whether the domain(s) of interaction among users (clinical matters vs. emotional and instrumental matters for instance) has an impact on predicting cancer stage by taking the different domains of interaction in account in the weight computation.

Finally, this work relies on a single, yet highly active and popular, forum. We would like to test our results on different breast cancer forums, but also on other disease-specific forums, where patients can be separated in clinically relevant groups.

Acknowledgments

We thank Phani Nivarthi for his help on data collection. This work is supported in part by a Google Research Award. Any opinions, findings, or conclusions are those of the authors, and do not necessarily reflect the views of the funding organization.

References

Erin Bantum and Jason Owen. 2009. Evaluating the validity of computerized content analysis programs for identification of emotional expression in cancer narratives. *Psychological Assessment*, 21(1):79–88.

Andrea Civan and Wanda Pratt. 2007. Threading together patient expertise. In *Proceedings of the AMIA Annual Symposium*, pages 140–144.

Brita Elvevaag, Peter Foltz, Mark Rosenstein, and Lynn DeLisi. 2010. An automated method to analyze language use in patients with schizophrenia and their first degree-relatives. *Journal of Neurolinguistics*, 23:270–284.

Gunther Eysenbach and James Till. 2001. Ethical issues in qualitative research on internet communities. *BMJ*, 323:1103–1105.

Kristi Graves, John Schmidt, Julie Bollmer, Michele Fejfar, Shelby Langer, Lee Blonder, and Michael Andrykowski. 2005. Emotional expression and emotional recognition in breast cancer survivors: A controlled comparison. *Psychology and Health*, 20(5):579–595.

Aria Haghighi. 2005. Robust textual inference via graph matching. In *Proceedings of the Conference on Empirical Methods in Natural Language Processing (EMNLP'05*, pages 387–394.

Mark Hall, Eibe Frank, Geoffrey Holmes, Bernhard Pfahringer, Peter Reutemann, and Ian H. Witten. 2009. The WEKA data mining software: An update. *SIGKDD Explorations*, 11(1).

Sanda Harabagiu, Finley Lacatusu, and Andrew Hickl. 2006. Answering complex questions with random walk models. In *Proceedings of SIGIR Conference (SIGIR'06)*, pages 220–227.

Pawel Jurczyk. 2007. Discovering authorities in question answer communities using link analysis. In *Proceedings of the ACM Conference on Information and Knowledge Management (CIKM'07)*.

Matthew Kreuter, David Farrell, Laura Olevitch, and Laura Brennan. 2000. *Tailoring health messages: customizing communication using computer technology*. Lawrence Erlbaum Associates.

Andrea Meier, Elizabeth Lyons, Gilles Frydman, Michael Forlenza, and Barbara Rimer. 2007. How cancer survivors provide support on cancer-related internet mailing lists. *Journal of Medical Internet Research*, 9(2):e12.

Zheng-Yu Niu, Dong-Hong Ji, and Chew Lim Tan. 2005. Word sense disambiguation using label propagation based semi-supervised learning. In *Proceedings of the ACL Conference (ACL'05)*, pages 395–402.

Thomas Oxman, Stanley Rosenberg, Paula Schnurr, and Gary Tucker. 1988. Diagnostic classification through content analysis of patient speech. *American Joural of Psychatry*, 145:464 468.

John Pestian, Pawel Matykiewicz, Jacqueline Grupp-Phelan, Sarah Arszman Lavanier, Jennifer Combs, and Robert Kowatch. 2008. Using natural language processing to classify suicide notes. In *Proceedings of BioNLP'08*, pages 96–97.

Usha Raghavan, Reka Albert, and Soundar Kumara. 2007. Near linear time algorithm to detect community structures in large-scale networks. *Physics Review*, page E 76 036106.

Linda Rozmovits and Sue Ziebland. 2004. What do patients with prostate or breast cancer want from an Internet site? a qualitative study of information needs. *Patient Education and Counseling*, 53:57–64.

Robert Schapire and Yoram Singer. 2000. BoosTexter: A boosting-based system for text categorization. *Machine Learning*, 39(2/3):135–168.

Cornelia van Uden-Kraan, Constance Drossaert, Erik Tall, Bret Shaw, Erwin Seydel, and Mart van de Laar. 2008. Empowering processes and outcomes of participation in online support groups for patients with breast cancer, arthritis, or fibromyalgia. *Qualitative Health Research*, 18(3):405–417.

An Exploration of Mining Gene Expression Mentions and their Anatomical Locations from Biomedical Text

Martin Gerner
Faculty of Life Sciences
University of Manchester
Manchester, UK
martin.gerner@postgrad.manchester.ac.uk

Goran Nenadic
School of Computer Science
University of Manchester
Manchester, UK
g.nenadic@manchester.ac.uk

Casey M. Bergman
Faculty of Life Sciences
University of Manchester
Manchester, UK
casey.bergman@manchester.ac.uk

Abstract

Here we explore mining data on gene expression from the biomedical literature and present Gene Expression Text Miner (GETM), a tool for extraction of information about the expression of genes and their anatomical locations from text. Provided with recognized gene mentions, GETM identifies mentions of anatomical locations and cell lines, and extracts text passages where authors discuss the expression of a particular gene in specific anatomical locations or cell lines. This enables the automatic construction of expression profiles for both genes and anatomical locations. Evaluated against a manually extended version of the BioNLP '09 corpus, GETM achieved precision and recall levels of 58.8% and 23.8%, respectively. Application of GETM to MEDLINE and PubMed Central yielded over 700,000 gene expression mentions. This data set may be queried through a web interface, and should prove useful not only for researchers who are interested in the developmental regulation of specific genes of interest, but also for database curators aiming to create structured repositories of gene expression information. The compiled tool, its source code, the manually annotated evaluation corpus and a search query interface to the data set extracted from MEDLINE and PubMed Central is available at http://getm-project.sourceforge.net/.

1 Introduction

With almost 2000 articles being published daily in 2009, the amount of available research literature in the biomedical domain is increasing rapidly. Currently, MEDLINE contains reference records for almost 20 million articles (with about 10 million abstracts), and PubMed Central (PMC) contains almost two million full-text articles. These resources store an enormous wealth of information, but are proving increasingly difficult to navigate and interpret. This is true both for researchers seeking information on a particular subject and for database curators aiming to collect and annotate information in a structured manner.

Text-mining tools aim to alleviate this problem by extracting structured information from unstructured text. Considerable attention has been given to some areas in text-mining, such as recognizing named entities (e.g. species, genes and drugs) (Rebholz-Schuhmann *et al.*, 2007; Hakenberg *et al.*, 2008; Gerner *et al.*, 2010) and extracting molecular relationships, e.g. protein-protein interactions (Donaldson *et al.*, 2003; Plake *et al.*, 2006; Chowdhary *et al.*, 2009). Many other areas of text mining in the biomedical domain are less mature, including the extraction of information about the expression of genes (Kim *et al.*, 2009). The literature contains a large amount of information about where and when genes are expressed, as knowledge about the expression of a gene is critical for understanding its function and has therefore often been reported as part of gene studies. Gene expression profiles from genome-wide studies are available in specialized databases such as the NCBI Gene Expression Omnibus (Barrett *et al.*, 2009) and FlyAtlas (Chintapalli *et al.*, 2007), but results on gene expression from smaller studies remain locked in the primary literature.

Previously, a number of data-mining projects have combined text-mining methods with structured genome-wide gene expression data in order

Proceedings of the 2010 Workshop on Biomedical Natural Language Processing, ACL 2010, pages 72–80,
Uppsala, Sweden, 15 July 2010.

to allow further interpretation of the gene expression data (Natarajan *et al.*, 2006; Fundel, 2007). However, only recently has interest in text-mining tools aimed at extracting gene expression profiles from primary literature started to grow. The 2009 BioNLP shared task (Kim *et al.*, 2009) aimed at extracting biological "events", where one of the event types was gene expression. For this event type, participants were asked to determine locations in text documents where authors discussed the expression of a gene or protein and extract a *trigger* keyword (e.g. "expression") and its associated gene *participant* (the gene whose expression is discussed). The group that achieved the highest accuracy on the "simple event" task (where gene expression extraction was included) achieved recall and precision levels of 64.2% and 77.5%, respectively (Björne *et al.*, 2009). A key limitation of the 2009 shared task was that all genes had been annotated prior to the beginning of the task, making it difficult to anticipate the accuracy of tools that do not rely on pre-annotated entities.

Biologists are interested not only in finding statements of gene expression events, but also in knowing where and when a gene is expressed. However, to the best of our knowledge, no effort has previously been made to extract and map the expression of genes to specific tissues and cell types (and vice versa) from the literature. Thus, we have taken preliminary steps to construct a software tool, named Gene Expression Text Miner (GETM), capable of extracting information about what genes are expressed and where they are expressed. An additional goal of this work is to apply this tool to the whole of MEDLINE and PMC, and make both the tool and the extracted data available to researchers.

We anticipate that the data extracted by GETM will provide researchers an overview about where a specific gene is expressed, or what genes are expressed in a specific anatomical location. Moreover, GETM will aid in the curation of gene expression databases by providing text passages and identifiers to database curators for verification.

2 Methods

An overview of the workflow of GETM is given in Figure 1. Articles are initially scanned for mentions of gene entities, anatomical entities and keywords indicating the discussion of gene expression (called *triggers* following BioNLP terminology, e.g. "expression" and "expressed in"). After the detection of the entities and triggers, abbreviations are detected and entities are grouped in the cases of enumerations. Finally, sentences are split and each sentence is processed in order to associate triggers with gene and anatomical entities. Each step is described below in more detail.

2.1 Named entity recognition and abbreviation detection

In order to extract information on the expression of genes and their anatomical locations, a key requirement is the accurate recognition and normalization (mapping the recognized terms to database identifiers) of both the genes and anatomical locations in question. In order to locate and identify gene names, we utilized GNAT (Hakenberg *et al.*, 2008), an inter-species gene name recognition software package. Among the gene name recognition tools capable of gene normalization, GNAT is currently showing the best accuracy (compared to the BioCreative corpora (Hirschman *et al.*, 2005; Morgan *et al.*, 2008)). The species identification component of GNAT, used to help disambiguate gene mentions across species, was performed by LINNAEUS (Gerner *et al.*, 2010).

In order to perform named entity recognition (NER) of anatomical locations, we investigated the use of various anatomical ontologies. A key challenge with these ontologies is that the terms

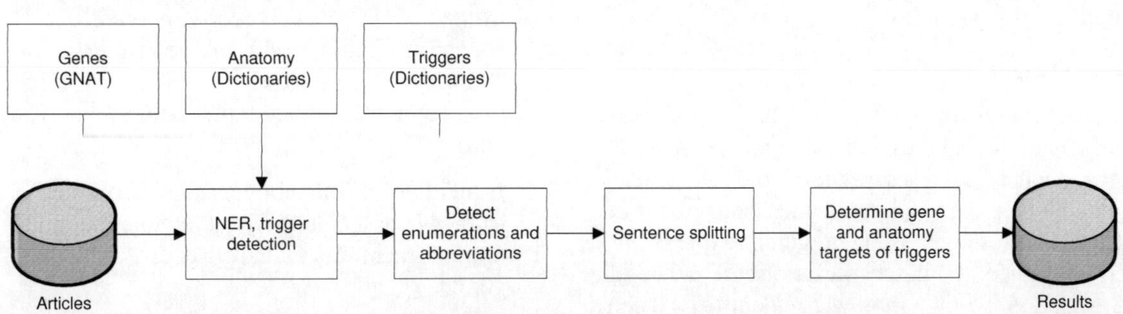

Figure 1. Schematic overview of the processing workflow of GETM.

vary significantly from one species to another. For example, fruit flies have wings while humans do not, and humans have fingers, while fruit flies do not. Efforts have been made in creating unified species-independent anatomical ontologies, such as Uberon (Haendel *et al.*, 2009; Mungall *et al.*, 2010). However, in preliminary experiments we found that the coverage of Uberon was not extensive enough for this particular application (data not shown), motivating us to instead use a combination of various species-specific anatomical ontologies hosted at the OBO Foundry (Smith *et al.*, 2007). These ontologies (n = 13) were chosen in order to cover terms from the main model organisms that are used in research (e.g. human, fruit fly, mouse, *Caenorhabditis elegans*) and a few larger groups of organisms such as e.g. amphibians and fungi. It is worth noting that the more general terms, such as e.g. "brain", are likely to match anatomical locations in other species as well. In total, the selected ontologies contain terms for 38,459 different anatomical locations.

We also utilized an ontology of cell lines (Romano *et al.*, 2009), containing terms for a total of 8,408 cell lines (ranging across 60 species), as cell lines can be viewed as biological proxies for the anatomical locations that gave rise to them. For example, the HeLa cell line was derived from human cervical cells, and the THP1 cell line was derived from human monocytes (Romano *et al.*, 2009).

The anatomical and cell line NER, utilizing the OBO Foundry and cell line ontologies, was performed using dictionary-matching methods similar to those employed by LINNAEUS (Gerner *et al.*, 2010).

After performing gene and anatomical NER on the document, abbreviations were detected (using the algorithm by Schwartz and Hearst (2003)) in order to allow the detection and markup of abbreviated entity names in the cases where the abbreviations do not exist in any of the ontologies that are used.

2.2 Trigger detection

The trigger keywords indicating that an author is discussing the expression of one or several genes, such as e.g. *"expression"* and *"expressed in"* were detected using a manually created list of regular expressions. The regular expressions were designed to match variations of a set of terms, listed below, that were identified when inspecting documents not used when building the gold-standard corpus (see Section 3.1).

The terms used to construct the trigger regular expressions were orthographical, morphological and derivational variations of "expression", "production" and "transcription". Descriptions of the level of expression were also considered for the different terms, such as "over-expression," "under-expression," "positively expressed," "negatively expressed," *etc.*

Each gene expression mention that has been extracted by GETM contains information about the trigger term used by the author, allowing researchers to look only at e.g. the "negative" mentions (where genes are e.g. "under-expressed" or "negatively expressed") or the "positive" mentions (where genes are e.g. "over-expressed").

2.3 Association of entities to the trigger

To help associate triggers with the correct gene and anatomical entities, articles were first split into sentences, allowing each sentence to be processed in turn. In order to reduce the number of false positives and preserve a high level of precision, any sentences that did not contain a trigger, at least one gene mention and at least one anatomical mention were ignored. For the sentences that did contain a combination of all three requirements (trigger, gene and anatomical mention), the following pattern- and distance-based rules were employed in order to associate each trigger with the correct gene and anatomical mention:

1. If there is only one gene mention and only one anatomical mention in the sentence, the trigger is associated with those mentions.
2. If there is one gene mention (G) and one anatomical mention (A) in the sentence such that they match one of the patterns "<G> is expressed in <A>", "expression of <G> in <A>", "<A> transcribes <G>" or "<A> produces <G>", the gene mention <G> and anatomical mention <A> are associated with the trigger (variations of the triggers, such as "over-expressed" and "negative expression" are considered as well). Additional gene or anatomical mentions that fall outside the pattern are ignored.
3. If neither of the above rules applies, the trigger is associated with the gene and anatomical mentions that are closest to the trigger.

For the purposes of these rules, an enumeration of several genes or anatomical locations was handled as if it was only a single mention. For example, Rule 1 might trigger even if there are several genes mentioned in the same sentence, as long as they are mentioned together as part of an enumeration.

In order to detect these enumerations, a rule-based algorithm for connecting enumerated gene and anatomical entity mentions (as in e.g. "... RelB and DC-CK1 gene expression ...") was also implemented. Being able to detect enumerations allowed the rules described above to recognize that a particular gene expression mention do not refer to only e.g. "RelB" or "DC-CK1", but both of them at the same time.

Each trigger was processed independently, allowing the potential extraction of multiple gene expression statements from a single sentence.

Initially, experiments were performed using stricter rules where only variations of Rule 2, requiring gene and anatomical mentions to conform to certain patterns, were used. However, recall was in these cases found to be extremely low (below 5%, data not shown). The current rules are more permissive, allowing higher recall.

The fact that the method requires a combination of a trigger, a gene and an anatomical location makes it susceptible to false negatives: if any one of them cannot be found by the NER or trigger detection methods, the whole combination is missed.

3 Evaluation

3.1 Extending the BioNLP shared task gold-standard corpus

In order to make a meaningful evaluation of the accuracy of text-mining applications, a gold-standard corpus, consisting of manually annotated mentions for a set of documents, is required. Previously, no such corpus existed that was suitable for this problem (providing annotations linked to mentions of both gene and anatomical locations). However, the BioNLP corpus (Ohta *et al.*, 2009) which is based on the GENIA corpus (Kim *et al.*, 2008), does contain annotations about gene expression. Annotations in the corpus contain trigger terms that are linked to genes (or gene products) where the authors discuss gene expression. However, anatomical locations have not been annotated in this corpus.

In order to allow evaluation of the accuracy of our software, we extended the annotations of gene expression events in part of the BioNLP corpus. Each gene expression entry in the corpus was linked to the anatomical location or cell line that the author mentioned. In cases where gene expression was only discussed generally without referring to expression in a particular location, no association to an anatomical location could be made (these entries were ignored during evaluation). Note that named entities were only linked to their locations in the text, not to unique database identifiers (such as Entrez Gene or OBO Foundry identifiers). Because of this, subsequent evaluation in this extended corpus is limited to the accuracy of recognition (locating the entities in the text), but not normalization (linking the entities to database identifiers).

In total, annotations for 150 abstracts (constituting the development set of the BioNLP corpus) were extended to also include anatomical locations. These abstracts contained 377 annotated gene expression events, of which 267 (71%) could be linked to anatomical locations. These results demonstrate that the majority of gene expression mentions include reference to an anatomical location. For a few cases where the author described the expression of a gene in several cell types, a single gene expression event gave rise to several distinct "entries" in the extended corpus, creating a total of 279 final gene expression entries that are linked to anatomical locations.

4 Results

In order to evaluate the accuracy of GETM, it was first run on the 150 abstracts in the gold-standard corpus, after which the extracted results were compared against the annotations of the corpus. GETM was also applied to the whole of MEDLINE and PMC, in order to extract a searchable and structured data set of gene expression mentions in published biomedical articles.

4.1 Accuracy

The gene expression mentions extracted by GETM from the corpus were compared against the manually created annotations in order to estimate the accuracy of the software. After inspecting the false positives and false negatives, we noted that a number of the false positives actually were correctly identified by our system and had been marked as false positives only because of incomplete annotations in the corpus. Because of this, all false positives were manually examined in order to determine the "correct" number of false positives. For one of the corrected expression mentions, two anatomical locations were enumerated, with GETM only locating one of them. This introduced both a new true positive (for the one that was recognized) and a new false negative (for the one that was not). The number of true positives, false positives, false negatives, precision and recall (before and after correction) are shown in Table 1.

	Original	**Corrected**
TP	53	67
FP	61 (p = 46.5%)	47 (p = 58.8%)
FN	214 (r = 19.8%)	215 (r = 23.8%)

Table 1. The number of true positives (TP), false positives (FP), false negatives (FN) and levels of precision (p) and recall (r) for GETM when compared against the gold-standard corpus.

4.2 Analysis of false negatives

In order to determine the causes of the relatively high number of false negatives, the gene entities, anatomical entities and triggers identified by GNAT and GETM were compared to the extended corpus, allowing us to determine the number of corpus entities that could not be found by the GNAT and GETM NER tools. An analysis was also performed in order to determine the number of corpus entries that were spread across several sentences, as any expression mentions spread over several sentences are missed by GETM.

The analysis results can be seen in Table 2, showing that virtually all false negatives are caused either by incomplete NER or multi-sentence entries. Only considering the NER, 68% of the gold-standard corpus annotated entries contain either a trigger (example FN: "detected"), gene (example FN: CD4) or anatomical location (example FN: "lymphoblastoid cells") that could not be located automatically. GETM was further limited by entities being spread across several sentences (n=66, 23.6%). In total, 74.3% of all entries could not be extracted correctly due to either incomplete NER, incomplete trigger detection or the entities being spread across multiple sentences. This limited recall to 25.7%, even if the rule-based method was working perfectly.

4.3 Analysis of false positives

Manual inspection of the false positives (after adjusting the false positives caused by incomplete annotations) allowed the identification of one clear cause: if the NER methods fail to recognize the entity associated with a manually annotated expression entry, but there are other entities (that have been recognized) in the sentence, those entities might be incorrectly associated with the trigger instead. For example, in the sentence "In conclusion, these data show that IL-10 induces *c-fos expression* in human *B-cells* by activation of tyrosine and serine/threonine kinases." (Bonig et al., 1996) (the correct entities and trigger are italicized), a correctly extracted entry would link c-fos to B-cells through the trigger expression. However, the gene NER component failed to recognize c-fos but did recognize IL-10, causing GETM to incorrectly associate IL-10 with B-cells. Either increasing the accuracy of the NER methods or performing deeper grammatical parsing could potentially reduce the number of false positives of this type. We note that the number of cases for this category (n = 15; 34%) only make up a minority of the total number of false positives, and the remainder have no easily identifiable common cause.

4.4 Application to MEDLINE and PMC documents

GETM was applied to the whole set of 10,240,192 MEDLINE entries from the 2010 baseline files that contain an abstract (many MEDLINE entries do not contain an abstract). From these abstracts, 578,319 statements could be extracted containing information about the expression of a gene and the location of this expression. In addition, GETM was also applied to the set of 186,616 full-text articles that make up the open-access portion of PMC (downloaded February 5th, 2010). The full-text articles allowed the extraction of 145,796 statements (an 18-fold increase in entries per article compared

Problem type	**Number of occurrences**
Trigger not found	58 (20.7%)
Gene not found	139 (49.6%)
Anatomical location not found	74 (26.4%)
Any of the entities or trigger not found	190 (67.9%)
Total number of entities not contained in a single sentence	66 (23.6%)
Total number of entities either not found or not in the same sentence	208 (74.3%)

Table 2. Breakdown of the causes for false negatives in GETM, relative to the total number of entries in the gold-standard corpus.

Gene	Anatomical location	Number of mentions
Interleukin 2	T cells	3511
Interferon, gamma	T cells	2088
CD4	T cells	1623
TNF	Macrophages	1596
TNF	Monocytes	1539
Interleukin 4	T cells	1323
Integrin, alpha M	Neutrophils	1063
Inteleukin 10	T cells	971
ICAM 1	Endothelial cells	964
Interleukin 2	Lymphocytes	876

Table 3. The ten most commonly mentioned combinations of genes and anatomical locations

to the MEDLINE abstracts). In total, 716,541 statements were extracted, not counting the abstracts in MEDLINE that also appear in PMC. Overall, the combined extracted information ranges across 25,525 different genes (the most common being *tumor necrosis factor (TNF superfamily, member 2)* in human) and 3,655 different anatomical locations (the most common being *T cells*). The most common combination concerns the expression of human *interleukin 2* in *T cells*. The 10 most commonly mentioned combinations of genes and anatomical locations are shown in Table 3. Overall, these results suggest that studies on gene expression in the field of mammalian immunology are the dominant signal in MEDLINE and PMC. The genes that were recognized and normalized range across 15 species, out of the 23 supported by GNAT (Hakenberg *et al.*, 2008). The most common species is human, as expected (Gerner *et al.*, 2010), followed by mouse, rat, chicken and cow.

The majority of statements were associated to anatomical locations from the OBO Foundry ontologies (n=649,819; 89.7%), while the remainder were associated to cell lines (n=74,294; 10.3%). This result demonstrates the importance of taking cell lines into account when attempting to identify anatomical entities.

Finally, a total of 73,721 (11.7%) of the statements extracted from MEDLINE contained either genes or anatomical locations that had been enumerated by the author, underscoring the importance of considering enumerations when designing text-mining algorithms.

4.5 Availability

GETM is available under an open source license, and researchers may freely download GETM, its source code and the extended gold-standard corpus from http://getm-project.sourceforge.net/. Also available on the web site is a search query interface where researchers may search for extracted gene expression entries relating to a particular gene, anatomical location or a combination of the two and view these in the context of the surrounding text.

5 Discussion

5.1 Overview of design philosophy

When constructing text-mining applications, a balance between precision (reflecting the relative number of false positives) and recall (reflecting the relative number of false negatives) is often used to optimize system performance. Accordingly, a measure which often is used to evaluate the accuracy of software is the F-score (the harmonic mean of the precision and recall). In this work, we have decided that rather than trying to maximize the F-score, we have put more focus on precision in order to ensure that the data extracted by GETM are of as high quality as possible. This typically leads to lower recall, causing the software to detect a relatively smaller number of relevant passages. Nonetheless, we believe that for this particular application, a smaller amount of data with higher quality would be more useful to curators and biologists than a larger amount of data that is less reliable.

5.2 Comparison with previous work

It is difficult to compare the precision and recall levels of GETM (at 58.8% and 23.8%, respectively) against other tools, as GETM is the first tool aiming to perform this particular task. The closest comparison that can be made is against the software evaluated in the BioNLP shared task (Kim *et al.*, 2009). However, software developed for the BioNLP shared task did not attempt to extract the anatomical location of gene expression mentions, nor did they need to identify the component entities involved. The tool with the highest accuracy for the simple event task (where gene expression extraction was included) showed precision and recall levels of 77.5% and 64.2%, respectively (Björne *et al.*, 2009). It is not clear how tools evaluated in the 2009 BioNLP shared task would perform if they identified entities themselves rather than using pre-annotated entities.

5.3 Limits on accuracy

When investigating the cause of the low level of recall, the main reason that emerged for the high number of false negatives was the high number of annotated entries that could not be automatically extracted due to at least one of the gene, anatomical or trigger mentions not being recognized. This fact underscores the importance of accurate NER for applications that rely on the extracted entity mentions, especially those that attempt to extract information from multiple entity types, like GETM. The results also demonstrate that NER, particularly in the case of gene name normalization, continues to pose a challenging problem. It is possible that using a combination of GNAT and other gene NER tools would improve the overall gene NER accuracy.

We further explored the effects of "perfect" gene NER on the accuracy of GETM by using the manual gene mention annotations supplied in the BioNLP corpus. Using the pre-annotated gene names increased the number of gene expression mentions recognized and the number of true positives, significantly improving recall (from 23.8% to 37.8%; data not shown). However, a number of additional false positives were also introduced, causing precision to decrease very slightly from 58.8% to 58.5% (data not shown). This demonstrates the complexity of gene expression mentions in text, indicating that a combination of accurate trigger detection, accurate NER (for both genes and anatomical locations) and deeper NLP methods are needed in order to accurately capture gene expression profiles in text.

A secondary cause of false negatives was a relatively high number of annotated corpus entries that spanned several sentences. The high proportion (23%) of multi-sentence entries in our extended corpus differs from previously reported results. For the event annotations in the BioNLP corpus, previous analyses showed that only 5% of all entries spanned several sentences (Björne *et al.*, 2009). This suggests that the mentions of anatomical locations are located outside of the "trigger sentence" more often than gene mentions or other entities in the BioNLP corpus.

6 Conclusions

In this paper, we have explored integrated mining of gene expression mentions and their anatomical locations from the literature and presented a new tool, GETM, which can be used to extract information about the expression of genes and where they are expressed from biomedical text. We have also extended part of a previously existing gold-standard corpus in order to allow evaluation of GETM. When evaluated against the gold-standard corpus, GETM performed with precision and recall levels of 58.8% and 23.8%, respectively.

The relatively low level of recall was primarily caused by incomplete recognition of individual entities, indicating that – in order to increase the recall of GETM – future work would primarily need to focus on increasing the accuracy of the NER methods. With more accurate NER, while increasing recall, the higher number of recognized entities is also expected to increase the number of false positives, causing a need for deeper NLP methods in order to preserve and increase the level of precision.

While having a low level of recall, GETM was nonetheless able to extract 716,541 statements from MEDLINE and PMC, constituting a large and potentially useful data set for researchers wishing to get an overview of gene expression for a particular gene or anatomical location. The high number of mentions extracted from MEDLINE can give an indication of the amount of data available in MEDLINE: if the recall on the BioNLP corpus is representative for MEDLINE as a whole, a tool with perfect accuracy might be able to extract almost 2.5 million entries.

The level of precision (p = 58.8%) will most likely not be high enough for researchers to rely on the extracted data for high-throughput bioinformatical experiments without some kind of verification. However, we believe that it nonetheless will be of high enough quality that researchers and curators will not feel inconvenienced by false positives, as currently the only alternatives are multi-word free text searches through PubMed or Google. Additionally, we provide an interface with the text context surrounding gene expression statements, making it easier for researchers to quickly locate relevant results.

In the future, we will aim to evaluate the normalization of entities detected by GETM in order to quantify the level to which the identifiers assigned to the entities are correct. In addition, both the gene and anatomical NER components could be improved in order to both reduce the number of false negatives and cover gene and anatomical terms for a wider range of species, beyond the common model organisms. We also believe that extending this work by utilizing deeper NLP methods (e.g. dependency parsers) could further improve the accuracy of GETM and related approaches to mining the abundance of data on gene expression in the biomedical literature.

Acknowledgements

We thank Jörg Hakenberg (Arizona State University) for providing access to GNAT. We also thank members of the Bergman and Nenadic groups for helpful comments and suggestions throughout the project, and three anonymous reviewers of this article for valuable comments that helped improve the manuscript. This work was funded by the University of Manchester and a BBSRC CASE studentship (to M.G.).

References

Barrett, T., Troup, D. B., Wilhite, S. E., Ledoux, P., Rudnev, D., Evangelista, C., Kim, I. F., Soboleva, A., Tomashevsky, M., Marshall, K. A., Phillippy, K. H., Sherman, P. M., Muertter, R. N. and Edgar, R. (2009). "NCBI GEO: archive for high-throughput functional genomic data." Nucleic Acids Res 37(Database issue): D885-90.

Björne, J., Heimonen, J., Ginter, F., Airola, A., Pahikkala, T. and Salakoski, T. (2009). "Extracting complex biological events with rich graph-based feature sets." In Proceedings of the Workshop on BioNLP: Shared Task Boulder, Colorado: 10-18.

Bonig, H., Korholz, D., Pafferath, B., Mauz-Korholz, C. and Burdach, S. (1996). "Interleukin 10 induced c-fos expression in human B cells by activation of divergent protein kinases." Immunol Invest 25(1-2): 115-28.

Chintapalli, V. R., Wang, J. and Dow, J. A. T. (2007). "Using FlyAtlas to identify better Drosophila models of human disease." Nature Genetics 39: 715-720.

Chowdhary, R., Zhang, J. and Liu, J. S. (2009). "Bayesian inference of protein-protein interactions from biological literature." Bioinformatics 25(12): 1536-42.

Donaldson, I., Martin, J., de Bruijn, B., Wolting, C., Lay, V., Tuekam, B., Zhang, S., Baskin, B., Bader, G. D., Michalickova, K., Pawson, T. and Hogue, C. W. (2003). "PreBIND and Textomy--mining the biomedical literature for protein-protein interactions using a support vector machine." BMC Bioinformatics 4: 11.

Fundel, K. (2007). Text Mining and Gene Expression Analysis Towards Combined Interpretation of High Throughput Data. Dissertation. Faculty of Mathematics, Computer Science and Statistics. München, Ludwig-Maximilians Universität.

Gerner, M., Nenadic, G. and Bergman, C. M. (2010). "LINNAEUS: a species name identification system for biomedical literature." BMC Bioinformatics 11: 85.

Haendel, M. A., Gkoutos, G. V., Lewis, S. E. and Mungall, C. J. (2009). "Uberon: towards a comprehensive multi-species anatomy ontology." In International Conference on Biomedical Ontology Buffalo, NY.

Hakenberg, J., Plake, C., Leaman, R., Schroeder, M. and Gonzales, G. (2008). "Inter-species normalization of gene mentions with GNAT." Bioinformatics 24(16): i126-i132.

Hirschman, L., Yeh, A., Blaschke, C. and Valencia, A. (2005). "Overview of BioCreAtIvE: critical assessment of information extraction for biology." BMC Bioinformatics 6 Suppl 1: S1.

Kim, J. D., Ohta, T., Pyysalo, S., Kano, Y. and Tsujii, J. i. (2009). "Overview of BioNLP'09 Shared Task on Event Extraction." In Proceedings of the Workshop on BioNLP: Shared Task, Boulder, Colorado, Association for Computational Linguistics: 1-9.

Kim, J. D., Ohta, T. and Tsujii, J. (2008). "Corpus annotation for mining biomedical events from literature." BMC Bioinformatics 9: 10.

Morgan, A., Lu, Z., Wang, X., Cohen, A., Fluck, J., Ruch, P., Divoli, A., Fundel, K., Leaman, R., Hakenberg, J., Sun, C., Liu, H., Torres, R., Krauthammer, M., Lau, W., Liu, H., Hsu, C., Schuemie, M., Cohen, K. and Hirschman, L. (2008). "Overview of BioCreative II gene normalization." Genome Biology 9(Suppl 2): S3.

Mungall, C. J., Gkoutos, G. V., Smith, C. L., Haendel, M. A., Lewis, S. E. and Ashburner, M. (2010). "Integrating phenotype ontologies across multiple species." Genome Biol 11(1): R2.

Natarajan, J., Berrar, D., Dubitzky, W., Hack, C., Zhang, Y., DeSesa, C., Van Brocklyn, J. R. and Bremer, E. G. (2006). "Text mining of full-text journal articles combined with gene expression analysis reveals a relationship between sphingosine-1-phosphate and invasiveness of a glioblastoma cell line." BMC Bioinformatics 7: 373.

Ohta, T., Kim, J.-D., Pyysalo, S., Wang, Y. and Tsujii, J. i. (2009). "Incorporating GENETAG-style annotation to GENIA corpus." In Workshop on BioNLP, Boulder, Colorado: 106-107.

Plake, C., Schiemann, T., Pankalla, M., Hakenberg, J. and Leser, U. (2006). "AliBaba: PubMed as a graph." Bioinformatics 22(19): 2444-5.

Rebholz-Schuhmann, D., Arregui, M., Gaudan, M., Kirsch, H. and Jimeno, A. (2007). "Text processing through Web services: Calling Whatizit." Bioinformatics 23(2): e237-e244.

Romano, P., Manniello, A., Aresu, O., Armento, M., Cesaro, M. and Parodi, B. (2009). "Cell Line Data Base: structure and recent improvements towards molecular authentication of human cell lines." Nucl. Acids Res. 37(suppl_1): D925-932.

Schwartz, A. S. and Hearst, M. A. (2003). "A simple algorithm for identifying abbreviation definitions in biomedical text." Pac Symp Biocomput: 451-62.

Smith, B., Ashburner, M., Rosse, C., Bard, J., Bug, W., Ceusters, W., Goldberg, L. J., Eilbeck, K., Ireland, A., Mungall, C. J., Leontis, N., Rocca-Serra, P., Ruttenberg, A., Sansone, S. A., Scheuermann, R. H., Shah, N., Whetzel, P. L. and Lewis, S. (2007). "The OBO Foundry: coordinated evolution of ontologies to support biomedical data integration." Nat Biotechnol 25(11): 1251-5.

Exploring Surface-level Heuristics for Negation and Speculation Discovery in Clinical Texts

Emilia Apostolova
DePaul University
Chicago, IL USA
emilia.aposto@gmail.com

Noriko Tomuro
DePaul University
Chicago, IL USA
tomuro@cs.depaul.edu

Abstract

We investigate the automatic identification of negated and speculative statements in biomedical texts, focusing on the clinical domain. Our goal is to evaluate the performance of simple, Regex-based algorithms that have the advantage of low computational cost, simple implementation, and do not rely on the accurate computation of deep linguistic features of idiosyncratic clinical texts. The performance of the NegEx algorithm with an additional set of Regex-based rules reveals promising results (evaluated on the BioScope corpus). Current and future work focuses on a bootstrapping algorithm for the discovery of new rules from unannotated clinical texts.

1 Motivation

Finding negated and speculative (hedging) statements is an important subtask for biomedical Information Extraction (IE) systems. The task of hedge detection is of particular importance in the sub-genre of clinical texts which tend to avoid unqualified negations or assertions.

Negation/Speculation discovery is typically broken down into two subtasks - discovering the negation/speculation cue (a phrase or a syntactic pattern) and establishing its scope. While a number of cue and scope discovery algorithms have been developed, high performing systems typically rely on machine learning and more involved feature creation. Deep linguistic feature creation could pose problems, as the idiosyncrasies of clinical texts often confuse off-the-shelf NLP feature generation tools (e.g. relying on proper punctuation and grammaticality). In addition, computationally expensive algorithms could pose problems for high-volume IE systems.

In contrast, simple Regex-based algorithms have demonstrated larger practical significance as they offer reasonable performance at a low development and computational cost. NegEx[1] (Chapman et al., 2001), a simple rule-based algorithm developed for the discovery of negation of findings and diseases in discharge summaries, has been implemented in a number of BioNLP systems, including Metamap[2], CaTIES[3], and Mayo Clinic's Clinical IE System (Savova et al., 2008). In NegEx, a list of phrases split into subsets are used to identify cues and their corresponding scopes (token widows preceding or following the cues).

2 Method

Negation/Speculation in general English could be expressed by almost any combination of morphologic, syntactic, semantic, and discourse-level means. However, the scientific 'dryness' of the biomedical genre and clinical texts in particular, limits language variability and simplifies the task. We evaluated the performance of the NegEx algorithm on the BioScope corpus (Szarvas et al., 2008). BioScope corpus statistics are shown in Tables 1 and 2.

Corpus Type	Sentences	Documents	Mean Document Size
Radiology Reports	7520	1954	3.85
Biological Full Papers	3352	9	372.44
Biological Paper Abstracts	14565	1273	11.44

Table 1: Statistics of the BioScope corpus. Document sizes represent number of sentences.

Corpus Type	Negation Cues	Speculation Cues	Negation	Speculation
Rad Reports	872	1137	6.6%	13.4%
Full Papers	378	682	13.76%	22.29%
Paper Abstracts	1757	2694	13.45%	17.69%

Table 2: The percentage of speculative sentences (last column) is larger than the percentage of negated sentences.

We first evaluated the performance of an unmodified version of the NegEx algorithm on the task of cue detection (Table 3). Without any tuning or modifications, NegEx performed well on identifying negation cues across all documents, achiev-

[1] http://code.google.com/p/negex/
[2] ©The National Library of Medicine
[3] http://caties.cabig.upmc.edu/Wiki.jsp?page=Home

Proceedings of the 2010 Workshop on Biomedical Natural Language Processing, ACL 2010, pages 81–82,
Uppsala, Sweden, 15 July 2010.

ing an F-score of 90% on the clinical texts. For the task of identifying speculation cues, we simply used the NegEx Conditional Possibility Phrase list (35 speculative cue phrases). The overall performance of this simplistic approach revealed poor results.

	TP	FP	FN	Precision	Recall	F-score
Negation						
Rad Reports	836	131	36	86.45	95.87	90.92
Full Papers	307	74	71	80.58	81.22	80.9
Paper Abstracts	1390	211	367	86.82	79.11	82.79
Speculation						
Rad Reports	62	1	1075	98.41	5.45	10.33
Full Papers	1	0	681	100.0	0.15	0.3
Paper Abstracts	0	5	2694	0.0	0.0	0

Table 3: NegEx performance on identifying Negation and Speculation Cues (non-exact boundary). (TP=true positive, FP=false positive, FN=false negative)

As shown in Figure 1, speculation cues exhibit wider variability and a rule matching only 35 phrases proved inefficient. To enrich the list of speculation cues, we used hedging cues from the FlySlip corpus of speculative sentences[4]. Without any synonym expansion or fine-tuning, the performance of speculation cue detection improved significantly as shown in Table 4, achieving an F-score of 86% on the clinical dataset[5].

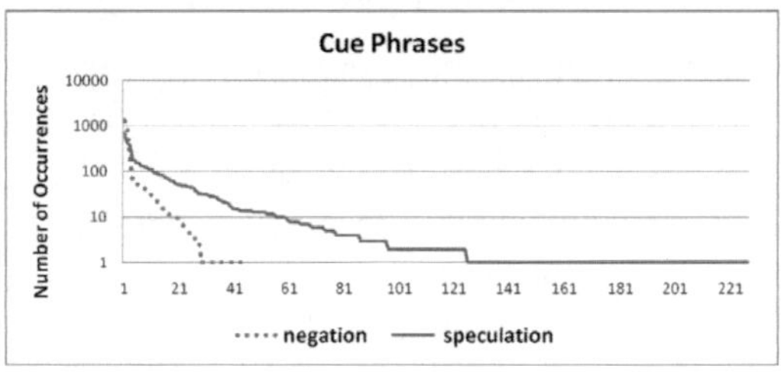

Figure 1: The number of occurrences (Y axis) of the 228 unique speculation cues and the 45 unique negation cues of the BioScope corpus (X axis).

Corpus	TP	FP	FN	Precision	Recall	F-score
Rad Reports	903	52	234	94.55	79.42	86.33
Full Papers	439	553	243	44.25	64.37	52.45
Paper Abstracts	1741	1811	953	49.01	64.63	55.75

Table 4: NegEx performance on identifying speculation cues (non-exact boundary) with the addition of the FlySlip hedging cues.

We next measured the performance of NegEx on scope detection. Newly introduced speculation cues from the FlySlip corpus were automatically classified into preceding or following their scope based the position of of their annotated 'topic'. Table 5 shows the results of scope identification.

3 Discussion

Our results show that a simple, surface-level algorithm could be sufficient for the task of negation and hedge detection in clinical texts. Using the NegEx algorithm and the FlySlip hedging corpus, without any modifications or additions, we were able to achieve an impressive F-score of 90.92% and 86.33% for negation and speculation cue discovery respectively[6]. We are currently expanding the set of speculation cues using an unannotated dataset of clinical texts and a bootstrapping algorithm (Medlock, 2008). The algorithm is based on the intuition that speculative cues tend to co-occur and this redundancy could be explored to probabilistically discover new cues from high-confidence existing ones. We are also exploring the discovery of degree of speculativeness (e.g. *very unlikely* vs *very likely*).

While NegEx performed well on the task of identifying negation scope (F-score 95.18), further work is needed on the discovery of speculation scopes (F-score 58.90). As hedging cues require a more fine-tuned set of rules, in future work we will evaluate linguistically motivated approaches (Kilicoglu and Bergler, 2008) for the creation of a set of surface-level speculation scope rules.

	TP	FP	FN	Precision	Recall	F-score
Negation						
Rad Reports	4003	267	140	94.12	97.61	95.18
Full Papers	2129	1835	525	54.45	80.12	64.01
Paper Abstracts	10049	6023	1728	63.04	85.13	72.31
Speculation						
Rad Reports	2817	1459	2471	65.87	53.27	58.90
Full Papers	3313	2372	2958	58.27	52.83	55.41
Paper Abstracts	17219	6329	9477	73.12	64.50	68.54

Table 5: NegEx performance on identifying scopes of correctly identified cues. Precision and recall are computed based on the number of correctly identified scope tokens excluding punctuation (i.e. number of tokens within cue scopes). Best results were achieved with no scope window size (i.e. using sentence boundaries).

References

W.W. Chapman, W. Bridewell, P. Hanbury, G.F. Cooper, and B.G. Buchanan. 2001. A simple algorithm for identifying negated findings and diseases in discharge summaries. *Journal of biomedical informatics*, 34(5):301–310.

H. Kilicoglu and S. Bergler. 2008. Recognizing speculative language in biomedical research articles: a linguistically motivated perspective. *BMC bioinformatics*, 9(Suppl 11):S10.

B. Medlock. 2008. Exploring hedge identification in biomedical literature. *Journal of Biomedical Informatics*, 41(4):636–654.

G.K. Savova, K. Kipper-Schuler, J.D. Buntrock, and C.G. Chute. 2008. UIMA-based Clinical Information Extraction System. In *Proc. UIMA for NLP Workshop. LREC.*

G. Szarvas, V. Vincze, R. Farkas, and J. Csirik. 2008. The BioScope corpus: annotation for negation, uncertainty and their scope in biomedical texts. In *Proceedings of the Workshop on Current Trends in Biomedical Natural Language Processing*, pages 38–45. Association for Computational Linguistics.

[4] http://www.wiki.cl.cam.ac.uk/rowiki/NaturalLanguage/FlySlip/Flyslip-resources

[5] To avoid fine-tuning cues on the corpus we did not set aside a training subset of the BioScope corpus for speculation cue enhancements and instead used an independent hedging corpus (FlySlip).

[6] The enhanced speculation cue phrase lists and a UIMA-based NegEx implementation are available upon request.

Disease Mention Recognition with Specific Features

Md. Faisal Mahbub Chowdhury [†‡] **and Alberto Lavelli** [‡]
[‡] Human Language Technology Research Unit, Fondazione Bruno Kessler, Trento, Italy
[†] ICT Doctoral School, University of Trento, Italy
{chowdhury,lavelli}@fbk.eu

Abstract

Despite an increasing amount of research on biomedical named entity recognition, there has been not enough work done on disease mention recognition. Difficulty of obtaining adequate corpora is one of the key reasons which hindered this particular research. Previous studies argue that correct identification of disease mentions is the key issue for further improvement of the disease-centric knowledge extraction tasks. In this paper, we present a machine learning based approach that uses a feature set tailored for disease mention recognition and outperforms the state-of-the-art results. The paper also discusses why a feature set for the well studied gene/protein mention recognition task is not necessarily equally effective for other biomedical semantic types such as diseases.

1 Introduction

The massive growth of biomedical literature volume has made the development of biomedical text mining solutions indispensable. One of the essential requirements for a text mining application is the ability to identify relevant entities, i.e. named entity recognition. Previous work on biomedical named entity recognition (BNER) has been mostly focused on gene/protein mention recognition. Machine learning (ML) based approaches for gene/protein mention recognition have already achieved a sufficient level of maturity (Torii et al., 2009). However, the lack of availability of adequately annotated corpora has hindered the progress of BNER research for other semantic types such as diseases (Jimeno et al., 2008; Leaman et al., 2009).

Correct identification of diseases is crucial for various disease-centric knowledge extraction tasks (e.g. drug discovery (Agarwal and Searls, 2008)). Previous studies argue that the most promising candidate for the improvement of disease related relation extraction (e.g. disease-gene) is the correct identification of concept mentions including diseases (Bundschus et al., 2008).

In this paper, we present a BNER system which uses a feature set specifically tailored for disease mention recognition. The system[1] outperforms other approaches evaluated on the Arizona Disease Corpus (AZDC) (more details in Section 5.1). One of the key differences between our approach and previous approaches is that we put more emphasis on the contextual features. We exploit syntactic dependency relations as well. Apart from the experimental results, we also discuss why the choice of effective features for recognition of disease mentions is different from that for the well studied gene/protein mentions.

The remaining of the paper is organized as follows. Section 2 presents a brief description of previous work on BNER for disease mention recognition. Then, Section 3 describes our system and Section 4 the feature set of the system. After that, Section 5 explains the experimental data, results and analyses. Section 6 describes the differences for the choice of feature set between diseases and genes/proteins. Finally, Section 7 concludes the paper with an outline of our future research.

2 Related Work

Named entity recognition (NER) is the task of locating boundaries of the entity mentions in a text and tagging them with their corresponding semantic types (e.g. person, location, gene and so on). Although several disease annotated corpora have been released in the last few years, they have been annotated primarily to serve the purpose of relation extraction and, for different reasons, they

[1]The source code of our system is available for download at http://hlt.fbk.eu/people/chowdhury/research

Proceedings of the 2010 Workshop on Biomedical Natural Language Processing, ACL 2010, pages 83–90,
Uppsala, Sweden, 15 July 2010.

are not suitable for the development of ML based disease mention recognition systems (Leaman et al., 2009). For example, the BioText (Rosario and Hearst, 2004) corpus has no specific annotation guideline and contains several inconsistencies, while PennBioIE (Kulick et al., 2004) is very specific to a particular sub-domain of diseases. Among other disease annotated corpora, EBI disease corpus (Jimeno et al., 2008) is not annotated with disease mention boundaries which makes it unsuitable for BNER evaluation for diseases. Recently, an annotated corpus, named as Arizona Disease Corpus (AZDC) (Leaman et al., 2009), has been released which has adequate and suitable annotation of disease mentions following specific annotation guidelines.

There has been some work on identifying diseases in clinical texts, especially in the context of CMC Medical NLP Challenge[2] and i2b2 Challenge[3]. However, as noted by Meystre et al. (2008), there are a number of reasons that make clinical texts different from texts of biomedical literature, e.g. composition of short, telegraphic phrases, use of implicit templates and pseudotables and so on. Hence, the strategies adopted for NER on clinical texts are not the same as the ones practiced for NER on biomedical literature.

As mentioned before, most of the work to date on BNER is focused on gene/protein mention recognition. State-of-the-art BNER systems are based on ML techniques such as conditional random fields (CRFs), support vector machines (SVMs) etc (Dai et al., 2009). These systems use either gene/protein specific features (e.g. Greek alphabet matching) or post-processing rules (e.g. extension of the identified mention boundaries to the left when a single letter with a hyphen precedes them (Torii et al., 2009)) which might not be as effective for other semantic type identification as they are for genes/proteins. There is a substantial agreement in the feature set that these systems use (most of which are actually various orthographical and morphological features).

Bundschus et al. (2008) have used a CRF based approach that uses typical features for gene/protein mention recognition (i.e. no feature tailoring for disease recognition) for disease, gene and treatement recognition. The work has been evaluated on two corpora which have been annotated with those entities that participate in disease-gene and disease-treatment relations. The reported results show F-measure for recognition of all the entities that participate in the relations and do not indicate which F-measure has been achieved specifically for disease recognition. Hence, the reported results are not applicable for comparison.

To the best of our knowledge, the only systematic experimental results reported for disease mention recognition in biomedical literature using ML based approaches are published by Leaman and Gonzalez (2008) and Leaman et al. (2009).[4] They have used a CRF based BNER system named BANNER which basically uses a set of orthographic, morphological and shallow syntactic features (Leaman and Gonzalez, 2008). The system achieves an F-score of 86.43 on the BioCreative II GM corpus[5] which is one of the best results for gene mention recognition task on that corpus.

BANNER achieves an F-score of 54.84 for disease mention recognition on the BioText corpus (Leaman and Gonzalez, 2008). However, as said above, the BioText corpus contains annotation inconsistencies[6]. So, the corpus is not ideal for comparing system performances. The AZDC corpus is much more suitable as it is annotated specifically for benchmarking of disease mention recognition systems. An improved version of BANNER achieves an F-score of 77.9 on AZDC corpus, which is the state of the art on ML based disease mention recognition in biomedical literature (Leaman et al., 2009).

3 Description of Our System

There are basically three stages in our approach – pre-processing, feature extraction and model training, and post-processing.

3.1 Pre-processing

At first, the system uses GeniaTagger[7] to tokenize texts and provide PoS tagging. After that, it corrects some common inconsistencies introduced by GeniaTagger inside the tokenized data (e.g. GeniaTagger replaces double inverted commas with

[2]http://www.computationalmedicine.org/challenge/index.php

[3]https://www.i2b2.org/NLP/Relations/Main.php

[4]However, there are some work on disease recognition in biomedical literature using other techniques such as morpho-syntactic heuristic based approach (e.g. MetaMap (Aronson, 2001)), dictionary look-up method and statistical approach (Névéol et al., 2009; Jimeno et al., 2008; Leaman et al., 2009).

[5]As mentioned in http://banner.sourceforge.net/

[6]http://biotext.berkeley.edu/data/dis_treat_data.html

[7]http://www-tsujii.is.s.u-tokyo.ac.jp/GENIA/tagger/

two single inverted commas). These PoS tagged tokized data are parsed using Stanford parser[8]. The dependency relations provided as output by the parser are used later as features. The tokens are further processed using the following generalization and normalization steps:

- each number (both integer and real) inside a token is replaced with *'9'*
- each token is further tokenized if it contains either punctuation characters or both digits and alphabetic characters
- all letters are changed to lower case
- all Greek letters (e.g. alpha) are replaced with *G* and Roman numbers (e.g. iv) with *R*
- each token is normalized using SPECIALIST lexicon tool[9] to avoid spelling variations

3.2 Feature extraction and model training

The features used by our system can be categorized into the following groups:

- general linguistic features (Table 1)
- orthographic features (Table 2)
- contextual features (Table 3)
- syntactic dependency features (Table 4)
- dictionary lookup features (see Section 4)

During dictionary lookup feature extraction, we ignored punctuation characters while matching dictionary entries inside sentences. If a sequence of tokens in a sentence matches an entry in the dictionary, the leftmost token of that sequence is labeled with B-DB and the remaining tokens of the sequence are labeled with I-DB. The label B-DB indicates the beginning of a dictionary match. If a token belongs to several dictionary matches, then all the other dictionary matches except the longest one are discarded.

The syntactic dependency features are extracted from the output of the parser while the general linguistic features are extracted directly from the preprocessed tokens. To collect the orthographic features, the original tokens inside the corresponding sentences are considered. The contextual features are derived using other extracted features and the original tokens.

Tokens are labeled with the corresponding disease annotations according to the IOB2 format. Our system uses Mallet (McCallum, 2002) to train a first-order CRF model. CRF is a state-of-the-art ML technique applied to a variety of text processing tasks including named entity recognition (Klinger and Tomanek, 2007) and has been successfully used by many other BNER systems (Smith et al., 2008).

3.3 Post-processing

Once the disease mentions are identified using the learned model, the following post-processing techniques are applied to reduce the number of wrong identifications:

- *Bracket mismatch correction:* If there is a mismatch of brackets in the identified mention, then the immediate following (or preceding) character of the corresponding mention is checked and included inside the mention if that character is the missing bracket. Otherwise, all the characters from the index where the mismatched bracket exists inside the identified mention are discarded from the corresponding mention.
- *One sense per discourse:* If any instance of a character sequence is identified as a disease mention, then all the other instances of that character sequence inside the same sentence are also annotated as disease mentions.
- *Short/long form annotation:* Using the algorithm of Schwartz and Hearst (2003), *"long form (short form)"* instances are detected inside sentences. If the short form is annotated as disease mention, then the long form is also annotated and vice versa.
- *Ungrammatical conjunction structure correction:* If an annotated mention contains comma (,) but there is no "and" in the following character sequence (from the character index of that comma) of that mention, then the annotation is splitted into two parts (at the index of the comma). Annotation of the original mention is removed and the splitted parts are annotated as two separate disease mentions.

[8]http://nlp.stanford.edu/software/lex-parser.shtml
[9]http://lexsrv3.nlm.nih.gov/SPECIALIST/index.html

- *Short and long form separation:* If both short and long forms are annotated in the same mention, then the original mention is discarded and the corresponding short and long forms are annotated separately.

4 Features for Disease Recognition

There are compelling reasons to believe that various issues regarding the well studied gene/protein mention recognition would not apply to the other semantic types. For example, Jimeno et al. (2008) argue that the use of disease terms in biomedical literature is well standardized, which is quite opposite for the gene terms (Smith et al., 2008).

After a thorough study and extensive experiments on various features and their possible combinations, we have selected a feature set specific to the disease mention identification which comprises features shown in Tables 1, 2, 4 and 3, and dictionary lookup features.

Feature name	Description
PoS	Part-of-speech tag
NormWord	Normalized token (see Section 3.1)
Lemma	Lemmatized form
charNgram	3 and 4 character n-grams
Suffix	2-4 character suffixes
Prefix	2-4 character prefixes

Table 1: General linguistic features for token_i

Feature name	Description
InitCap	Is initial letter capital
AllCap	Are all letters capital
MixCase	Does contain mixed case letters
SingLow	Is a single lower case letter
SingUp	Is a single upper case letter
Num	Is a number
PuncChar	Punctuation character (if token_i is a punctuation character)
PrevCharAN	Is previous character alphanumeric

Table 2: Orthographic features for token_i

Feature name	Description
$\text{Bi-gram}_{k,k+1}$ for $i-2 \leq k < i+2$	Bi-grams of normalized tokens
$\text{Tri-gram}_{k,k+1,k+2}$ for $i-2 \leq k < i+2$	Tri-grams of normalized tokens
$\text{CtxPoS}_{k,k+1}$ for $i \leq k < i+2$	Bi-grams of token PoS
$\text{CtxLemma}_{k,k+1}$ for $i \leq k < i+2$	Bi-grams of lemmatized tokens
$\text{CtxWord}_{k,k+1}$ for $i-2 \leq k < i+2$	Bi-grams of original tokens
Offset conjunctions	Extracted by Mallet from features in the range from token_{i-1} to token_{i+1}

Table 3: Contextual features for token_i

Feature name	Description
dobj	Target token(s) to which token_i is a direct object
iobj	Target token(s) to which token_i is an indirect object
nsubj	Target token(s) to which token_i is an active nominal subject
nsubjpass	Target token(s) to which token_i is a passive nominal subject
nn	Target token(s) to which token_i is a noun compound modifier

Table 4: Syntactic dependency features for token_i. For example, in the sentence "Clinton defeated Dole", "Clinton" is the *nsubj* of the *target token* "defeated".

Like Leaman et al. (2009), we have created a dictionary with the instances of the following nine of the twelve UMLS semantic types from the semantic group "DISORDER"[10] from UMLS Metathesaurus (Bodenreider, 2004): (i) *disease or syndrome*, (ii) *neoplastic process*, (iii) *congenital abnormality*, (iv) *acquired abnormality*, (v) *experimental model of disease*, (vi) *injury or poisoning*, (vii) *mental or behavioral dysfunction*, (viii) *pathological function* and (ix) *sign or symptom*. We have not considered the other three semantic types (*findings*, *anatomical abnormality* and *cell or molecular Dysfunction*) since these three types have not been used during the annotation of Arizona Disease Corpus (AZDC) which we have used in our experiments.

Previous studies have shown that dictionary lookup features, i.e. name matching against a

[10]http://semanticnetwork.nlm.nih.gov/SemGroups/

dictionary of terms, often increase recall (Torii et al., 2009; Leaman et al., 2009). However, an unprocessed dictionary usually does not boost overall performance (Zweigenbaum et al., 2007). So, to reduce uninformative lexical differences or spelling variations, we generalize and normalize the dictionary entries using exactly the same steps followed for the pre-processing of sentences (see Section 3.1).

To reduce chances of false and unlikely matches, any entry inside the dictionary having less than 3 characters or more than 10 tokens is discarded.

5 Experiments

5.1 Data

We have done experiments on the recently released Arizona Disease Corpus (AZDC)[11] (Leaman et al., 2009). The corpus has detailed annotations of diseases including UMLS codes, UMLS concept names, possible alternative codes, and start and end points of disease mentions inside the corresponding sentences. These detailed annotations make this corpus a valuable resource for evaluating and benchmarking text mining solutions for disease recognition. Table 5 shows various characteristics of the corpus.

Item name	Total count
Abstracts	793
Sentences	2,783
Total disease mentions	3,455
Disease mentions without overlaps	3,093
Disease mentions with overlaps	362

Table 5: Various characteristics of AZDC.

For the overlapping annotations, (e.g. "endometrial and ovarian cancers" and "ovarian cancers") we have considered only the larger annotations in our experiments. There remain 3,224 disease mentions after resolving overlaps according to the aforementioned criterion. We have observed minor differences in some statistics of the AZDC reported by Leaman et al. (2009) with the statistics of the downloadable version[12] (Table 5). However, these differences can be considered negligible.

5.2 Results

We follow an experimental setting similar to the one in Leaman et al. (2009) so that we can compare our results with that of the BANNER system. We performed 10-fold cross validation on AZDC in such a way that all sentences of the same abstract are included in the same fold. The results of all folds are averaged to obtain the final outcome. Table 6 shows the results of the experiments with different features using the exact matching criterion.

As we can see, our approach achieves significantly higher result than that of BANNER. Initially, with only the general linguistic and orthographic features the performance is not high. However, once the contextual features are used, there is a substantial improvement in the result. Note that BANNER does not use contextual features. In fact, the use of contextual features is also quite limited in other BNER systems that achieve high performance for gene/protein identification (Smith et al., 2008).

Dictionary lookup features provide a very good contribution in the outcome. This supports the argument of Jimeno et al. (2008) that the use of disease terms in biomedical literature is well standardized. Post-processing and syntactic dependency features also increase some performance.

We have done statistical significance tests for the last four experimental results shown in Table 6. For each of such four experiments, the immediate previous experiment is considered as the baseline. The tests have been performed using the approximate randomization procedure (Noreen, 1989). We have set the number of iterations to 1,000 and the confidence level to 0.01. According to the tests, the contributions of contextual features and dictionary lookup features are statistically significant. However, we have found that the contributions of post-processing rules and syntactic dependency features are statistically significant only when the confidence level is 0.2 or more. Since AZDC consists of only 2,783 sentences, we can assume that the impact of post-processing rules

[11] Downloaded from http://diego.asu.edu/downloads/AZDC/ at 5-Feb-2009

[12] Note that *"Disease mentions (total)"* in the paper of Leaman et al. (2009) actually refers to the *total disease mentions after overlap resolving* (Robert Leaman, personal communication). One other thing is, Leaman et al. (2009) mention 794 abstracts, 2,784 sentences and 3,228 (overlap resolved) disease mentions in the AZDC. But in our downloaded version of AZDC, there is 1 abstract missing (i.e. total 793 abstracts instead of 794). As a result, there is 1 less sentence and 4 less (overlap resolved) disease mentions than the originally reported numbers.

and syntactic dependency features has been not so significant despite of some performance improvement.

5.3 Error analysis

One of the sources of errors is the annotations having conjunction structures. There are 94 disease mentions in the data which contain the word "and". The boundaries of 11 of them have been wrongly identified during experiments, while 39 of them have been totally missed out by our system. Our system also has not performed well for disease annotations that have some specific types of prepositional phrase structures. For example, there are 80 disease annotations having the word "of" (e.g. "deficient activity of acid beta-glucosidase GBA"). Only 28 of them are correctly annotated by our system. The major source of errors, however, concerns abbreviated disease names (e.g. "PNH"). We believe one way to reduce this specific error type is to generate a list of possible abbreviated disease names from the long forms of disease names available in databases such as UMLS Metathesaurus.

6 Why Features for Diseases and Genes/Proteins are not the Same

Many of the existing BNER systems, which are mainly tuned for gene/protein identification, use features such as token shape (also known as word class and brief word class (Settles, 2004)), Greek alphabet matching, Roman number matching and so forth. As mentioned earlier, we have done extensive experiments with various feature combinations for the selection of disease specific features. We have observed that many of the features used for gene/protein identification are not equally effective for disease identification. Table 7 shows some of the results of those experiments.

This observation is reasonable because gene/protein names are much more complex than entities such as diseases. For example, they often contain punctuation characters (such as parentheses or hyphen), Greek alphabets and digits which are unlikely in disease names. Ideally, the ML algorithm itself should be able to utilize information from only the useful features and ignore the others in the feature set. But practically, having non-informative features often mislead the model learning. In fact, several surveys have argued that the choice of features matter at least as much as the choice of the algorithm if not more (Nadeau and Sekine, 2007; Zweigenbaum et al., 2007).

One of the interesting trends in gene/protein mention identification is to not utilize syntactic dependency relations (with the exception of Vlachos (2007)). Gene/protein names in biomedical literature are often combined (i.e. without being separated by space characters) with other characters which do not belong to the corresponding mentions (e.g. *p53*-mediated). Moreover, as mentioned before, gene/protein mentions commonly have very complex structures (e.g. *PKR(1-551)K64E/K296R* or *RXRalphaF318A*). So, it is a common practice to tokenize gene/proten names adopting an approach that split tokens as much as possible to extract effective features (Torii et al., 2009; Smith et al., 2008). But while the extensive tokenization boosts performance, it is often difficult to correctly detect dependency relations for the tokens of the gene/protein names in the sentences where they appear. As a result, use of the syntactic dependency relations is not beneficial in such approaches.[13] In comparison, disease mentions are less complex. So, the identified dependencies for disease mentions are more reliable and hence may be usable as potential features (refer to our experimental results in Table 6).

The above mentioned issues are some of the reasons why a feature set for the well studied gene/protein focused BNER approaches is not necessarily suitable for other biomedical semantic types such as diseases.

7 Conclusion

In this paper, we have presented a single CRF classifier based BNER approach for disease mention identification. The feature set is constructed using disease-specific contextual, orthographic, general linguistic, syntactic dependency and dictionary lookup features. We have evaluated our approach on AZDC corpus. Our approach achieves significantly higher result than BANNER which is the current state-of-the-art ML based approach for disease mention recognition. We have also explained why the choice of features for the well studied gene/protein does not apply for other semantic types such as diseases.

[13] We have done some experiments on Biocreative II GM corpus with syntactic dependency relations of the tokens, which are not reported in this paper, and the results support our argument.

System	Note	Precision	Recall	F-score
BANNER	(Leaman et al., 2009)	80.9	75.1	**77.9**
Our system	Using general linguistic and orthographic features	74.90	71.01	72.90
Our system	After adding contextual features	82.15	75.81	78.85
Our system	After adding post-processing	81.57	76.61	79.01
Our system	After adding syntactic dependency features	82.07	76.66	79.27
Our system	After adding dictionary lookup features	83.21	79.06	**81.08**

Table 6: 10-fold cross validation results using exact matching criteria on AZDC.

Experiment	Note	Precision	Recall	F-score
(i)	Using general linguistic, orthographic and contextual features	82.15	75.81	78.85
(ii)	After adding *WC* and *BWC* features in *(i)*	82.08	75.57	78.69
(iii)	After adding *IsGreekAlphabet*, *HasGreekAlphabet* and *IsRomanNumber* features in *(i)*	82.10	75.69	78.76

Table 7: Experimental results of our system after using some of the gene/protein specific features for disease mention recognition on AZDC. Here, *WC* and *BWC* refer to the "word class" and "brief word class" respectively.

Future work includes implementation of disease mention normalization (i.e. associating a unique identifier for each disease mention). We also plan to improve our current approach by including more contextual features and post-processing rules.

Acknowledgments

This work was carried out in the context of the project "eOnco - Pervasive knowledge and data management in cancer care". The authors would like to thank Robert Leaman for sharing the settings of his experiments on AZDC.

References

Agarwal, P., Searls, D. 2008. Literature mining in support of drug discovery. *Brief Bioinform*, 9(6):479–492.

Aronson, A. 2001. Effective mapping of biomedical text to the UMLS Metathesaurus: the MetaMap program. In *Proceedings AMIA Symposium*, pages 17–21.

Bodenreider, O. 2004. The Unified Medical Language System (UMLS): integrating biomedical terminology. *Nucleic Acids Research*, 32(suppl_1):D267–270, January.

Bundschus, M., Dejori, M., Stetter, M., Tresp, V., Kriegel, H. 2008. Extraction of semantic biomedical relations from text using conditional random fields. *BMC Bioinformatics*, 9:207.

Dai, H., Chang, Y., Tsai, R., Hsu, W. 2009. New challenges for biological text-mining in the next decade. *Journal of Computer Science and Technology*, 25(1):169–179.

Jimeno, A., Jimnez-Ruiz, E., Lee, V., Gaudan, S., Berlanga, R., Rebholz-Schuhmann, D. 2008. Assessment of disease named entity recognition on a corpus of annotated sentences. *BMC Bioinformatics*, 9(S-3).

Klinger, R., Tomanek, K. 2007. Classical Probabilistic Models and Conditional Random Fields. Technical Report TR07-2-013, Department of Computer Science, Dortmund University of Technology, December.

Kulick, S., Bies, A., Liberman, M., Mandel, M., McDonald, R., Palmer, M., Schein, A., Ungar, L. 2004. Integrated annotation for biomedical information extraction. In *Proceedings of HLT/NAACL 2004 BioLink Workshop*, pages 61–68.

Leaman, R., Gonzalez, G. 2008. Banner: An executable survey of advances in biomedical named entity recognition. In *Proceedings of Pacific Symposium on Biocomputing*, volume 13, pages 652–663.

Leaman, R., Miller, C., Gonzalez, G. 2009. Enabling recognition of diseases in biomedical text with machine learning: Corpus and benchmark. In *Proceedings of the 3rd International Symposium on Languages in Biology and Medicine*, pages 82–89.

McCallum, A. 2002. Mallet: A machine learning for language toolkit. http://mallet.cs.umass.edu,.

Meystre, S., Savova, G., Kipper-Schuler, K., Hurdle, J. 2008. Extracting information from textual documents in the electronic health record: a review of recent research. *IMIA Yearbook of Medical Informatics*, pages 128–44.

Névéol, A., Kim, W., Wilbur, W., Lu, Z. 2009. Exploring two biomedical text genres for disease recognition. In *Proceedings of the BioNLP 2009 Workshop*, pages 144–152, June.

Nadeau, D., Sekine, S. 2007. A survey of named entity recognition and classification. *Linguisticae Investigationes*, 30(1):3–26.

Noreen, E.W. 1989. *Computer-Intensive Methods for Testing Hypotheses: An Introduction.* Wiley-Interscience.

Rosario, B., Hearst, M. 2004. Classifying semantic relations in bioscience texts. In *Proceedings of the 42nd Meeting of the Association for Computational Linguistics (ACL'04).*

Schwartz, A., Hearst, M. 2003. A simple algorithm for identifying abbreviation definitions in biomedical text. In *Proceedings of Pacific Symposium on Biocomputing*, pages 451–62.

Settles, B. 2004. Biomedical named entity recognition using conditional random fields and rich feature sets. In *Proceedings of the International Joint Workshop on Natural Language Processing in Biomedicine and its Applications*, pages 104–107.

Smith, L., Tanabe, L., Ando, R., Kuo, C., et al. 2008. Overview of BioCreative II gene mention recognition. *Genome Biology*, 9(Suppl 2).

Torii, M., Hu, Z., Wu, C., Liu, H. 2009. Biotagger-GM: a gene/protein name recognition system. *Journal of the American Medical Informatics Association : JAMIA*, 16:247–255.

Vlachos, A. 2007. Tackling the BioCreative2 gene mention task with conditional random fields and syntactic parsing. In *Proceedings of the 2nd BioCreative Challenge Evaluation Workshop*, pages 85–87.

Zweigenbaum, P., Demner-Fushman, D., Yu, H., Cohen, K. 2007. Frontiers of biomedical text mining: current progress. *Brief Bioinform*, 8(5):358–375.

Extraction of Disease-Treatment Semantic Relations from Biomedical Sentences

Oana Frunza and Diana Inkpen
School of Information Technology and Engineering
University of Ottawa Ottawa, ON, Canada, K1N 6N5
{ofrunza,diana}@site.uottawa.ca

Abstract

This paper describes our study on identifying semantic relations that exist between diseases and treatments in biomedical sentences. We focus on three semantic relations: *Cure*, *Prevent*, and *Side Effect*. The contributions of this paper consists in the fact that better results are obtained compared to previous studies and the fact that our research settings allow the integration of biomedical and medical knowledge. We obtain 98.55% F-measure for the *Cure* relation, 100% F-measure for the *Prevent* relation, and 88.89% F-measure for the *Side Effect* relation.

1 Introduction

Research in the fields of life-science and biomedical domain has been the focus of the Natural Language Processing (NLP) and Machine Learning (ML) community for some time now. This trend goes very much inline with the direction the medical healthcare system is moving to: the electronic world. The research focus of scientists that work in the filed of computational linguistics and life science domains also followed the trends of the medicine that is practiced today, an Evidence Based Medicine (EBM). This new way of medical practice is not only based on the experience a healthcare provider acquires as time passes by, but on the latest discoveries as well. We live in an information explosion era where it is almost impossible to find that piece of relevant information that we need. With easy and cheep access to disk-space we sometimes even find challenging to find our stored local documents. It should come to no surprise that the global trend in domains like biomedicine and not only is to rely on technology to identify and upraise information. The amount of publications and research that is indexed in the life-science domain grows almost exponentially (Hunter and Cohen (2006) making the task of finding relevant information, a hard and challenging task for NLP research.

The search for information in the life-science domain is not only the focus of researchers that work in these fields, but the focus of laypeople as well. Studies reveal that people are searching the web for medical-related articles to be better informed about their health. Ginsberg *et al.* (2009) show how a new outbreak of the influenza virus can be detected from search engine query data.

The aim of this paper is to show which NLP and ML techniques are suitable for the task of identifying semantic relations between diseases and treatments in short biomedical texts. The value of our work stands in the results we obtain and the new feature representation techniques.

2 Related Work

The most relevant work for our study is the work of Rosario and Hearst (2004). The authors of this paper are the ones that created and distributed the data set used in our research. The data set is annotated with disease and treatments entities and with 8 semantic relations between diseases and treatments. The main focus of their work is on entity recognition – the task of identifying entities, diseases and treatments in biomedical text sentences. The authors use Hidden Markov Models and maximum entropy models to perform both the task of entity recognition and of relation discrimination. Their representation techniques are based on words in context, part-of-speech information, phrases, and terms from MeSH[1], a medical lexical knowledge-base. Compared to previous work, our research is focused

[1] http://www.nlm.nih.gov/mesh/meshhome.html

Proceedings of the 2010 Workshop on Biomedical Natural Language Processing, ACL 2010, pages 91–98,
Uppsala, Sweden, 15 July 2010. ©2010 Association for Computational Linguistics

on different representation techniques, different classification models, and most importantly in obtaining improved results without using the annotations of the entities (new data will not have them). In previous research, the best results were obtained when the entities involved in the relations were identified and used as features.

The biomedical literature contains a wealth of work on semantic relation extraction, mostly focused on more biology-specific tasks: *subcellular-location* (Craven 1999), *gene-disorder* association (Ray and Craven 2001), and *diseases and drugs* relations (Srinivasan and Rindflesch 2002, Ahlers et *al.*, 2007).

Text classification techniques combined with a Naïve Bayes classifier and relational learning algorithms are methods used by Craven (1999). Hidden Markov Models are used in Craven (2001), but similarly to Rosario and Hearst (2004), the research focus was entity recognition.

A context based approach using MeSH term co-occurrences are used by Srinivasan and Rindflesch (2002) for relationship discrimination between diseases and drugs.

A lot of work is focused on building rules used to extract relation. Feldman et *al.* (2002) use a rule-based system to extract relations that are focused on genes, proteins, drugs, and diseases. Friedman et *al.* (2001) go deeper into building a rule-based system by hand-crafting a semantic grammar and a set of semantic constraints in order to recognize a range of biological and molecular relations.

3 Task and Data Sets

Our task is focused on identifying disease-treatment relations in sentences. Three relations: *Cure*, *Prevent*, and *Side Effect*, are the main objective of our work. We are tackling this task by using techniques based on NLP and supervised ML techniques. We decided to focus on these three relations because these are the ones that are better represented in the original data set and in the end will allow us to draw more reliable conclusions. Also, looking at the meaning of all relations in the original data set, the three that we focus on are the ones that could be useful for wider research goals and are the ones that really entail relations between two entities. In the supervised ML settings the amount of training data is a factor that influences the performance; support for this stands not only in the related work performed on the same data set, but in the research literature as well. The aim of this paper is to focus on few relations of interest and try to identify what predictive model and what representation techniques bring the best results of identifying semantic relations in short biomedical texts. We mostly focused on the value that the research can bring, rather than on an incremental research.

As mentioned in the previous section, the data set that we use to run our experiments is the one of Rosario and Hearst (2004). The entire data set is collected from Medline[2] 2001 abstracts. Sentences from titles and abstracts are annotated with entities and with 8 relations, based only on the information present in a certain sentence. The first 100 titles and 40 abstracts from each of the 59 Medline 2001 files were used for annotation. Table 1, presents the original data set, as published in previous research. The numbers in parenthesis represent the training and test set sizes.

Relationship	Definition and Example
Cure 810 (648, 162)	TREAT cures DIS *Intravenous immune globulin for recurrent spontaneous abortion*
Only DIS 616 (492, 124)	TREAT not mentioned *Social ties and susceptibility to the common cold*
Only TREAT 166 (132, 34)	DIS not mentioned *Flucticasome propionate is safe in recommended doses*
Prevent 63 (50, 13)	TREAT prevents the DIS *Statins for prevention of stroke*
Vague 36 (28, 8)	Very unclear relationship *Phenylbutazone and leukemia*
Side Effect 29 (24, 5)	DIS is a result of a TREAT *Malignant mesodermal mixed tumor of the uterus following irradiation*
NO Cure 4 (3, 1)	TREAT does not cure DIS *Evidence for double resistance to permethrin and malathion in head lice*
Total relevant: 1724 (1377, 347)	
Irrelevant 1771 (1416, 355)	Treat and DIS not present *Patients were followed up for 6 months*
Total: 3495 (2793, 702)	

Table 1. Original data set.

From this original data set, the sentences that are annotated with *Cure*, *Prevent*, *Side Effect*, *Only DIS*, *Only TREAT*, and *Vague* are the ones that used in our current work. While our main focus is on the *Cure*, *Prevent*, and *Side Effect*, we also run experiments for all relations such that a direct comparison with the previous work is done.

[2] http://medline.cos.com/

Table 2 describes the data sets that we created from the original data and used in our experiments. For each of the relations of interest we have 3 labels attached: *Positive*, *Negative*, and *Neutral*. The *Positive* label is given to sentences that are annotated with the relation in question in the original data; the *Negative* label is given to the sentences labeled with *Only DIS* and *Only TREAT* classes in the original data; *Neutral* label is given to the sentences annotated with *Vague* class in the original data set.

		Train	
Relation	**Positive**	**Negative**	**Neutral**
Cure	554	531	25
Prevent	42	531	25
SideEffect	20	531	25
		Test	
Relation	**Positive**	**Negative**	**Neutral**
Cure	276	266	12
Prevent	21	266	12
SideEffect	10	266	12

Table 2. Our data sets[3].

4 Methodology

The experimental settings that we follow are adapted to the domain of study (we integrate additional medical knowledge), yielding for the methods to bring improved performance.

The challenges that can be encountered while working with NLP and ML techniques are: finding the suitable model for prediction – since the ML field offers a suite of predictive models (algorithms), the task of finding the suitable one relies heavily on empirical studies and knowledge expertise; and finding the best data representation – identifying the right and sufficient features to represent the data is a crucial aspect. These challenges are addressed by trying various predictive algorithms based on different learning techniques, and by using various textual representation techniques that we consider suitable.

The task of identifying the three semantic relations is addressed in three ways:

Setting 1: build three models, each focused on one relation that can distinguish sentences that contain the relation – *Positive* label, from other sentences that are neutral – *Neutral* label, and from sentences that do not contain relevant information – *Negative* label;

Setting 2: build three models, each focused on one relation that can distinguish sentences that contain the relation from sentences that do not contain any relevant information. This setting is similar to a two-class classification task in which instances are labeled either with the relation in question – *Positive* label, or with non-relevant information – *Negative* label;

Setting 3: build one model that distinguishes the three relations – a three-way classification task where each sentence is labeled with one of the semantic relations, using the data with all the *Positive* labels.

The first set of experiments is influenced by previous research done by Koppel and Schler (2005). The authors claim that for polarity learning "neutral" examples help the learning algorithms to better identify the two polarities. Their research was done on a corpus of posts to chat groups devoted to popular U.S. television and posts to shopping.com's product evaluation page.

As classification algorithms, a set of 6 representative models: decision-based models (Decision trees – J48), probabilistic models (Naïve Bayes and complement Naïve Bayes (CNB), which is adapted for imbalanced class distribution), adaptive learning (AdaBoost), linear classifier (support vector machine (SVM) with polynomial kernel), and a classifier, ZeroR, that always predicts the majority class in the training data used as a baseline. All classifiers are part of a tool called Weka[4].

As representation technique, we rely on features such as the words in the context, the noun and verb-phrases, and the detected biomedical and medical entities. In the following subsections, we describe all the representation techniques that we use.

4.1 Bag-of-words representation

The bag-of-words (BOW) representation is commonly used for text classification tasks. It is a representation in which the features are chosen among the words that are present in the training data. Selection techniques are used in order to identify the most suitable words as features. After the feature space is identified, each training and test instance is mapped into this feature representation by giving values to each feature for a certain instance. Two feature value representations are the most commonly used for the BOW representation: binary feature values – the value

[3] The number of sentences available for download is not the same as the ones from the original data set, published in Rosario and Hearst ('04).

[4] http://www.cs.waikato.ac.nz/ml/weka/

of a feature is 1 if the feature is present in the instance and 0 otherwise, or frequency feature values – the feature value is the number of times it appears in an instance, or 0 if it did not appear.

Taking into consideration the fact that an instance is a sentence, the textual information is relatively small. Therefore a frequency value representation is chosen. The difference between a binary value representation and a frequency value representation is not always significant, because sentences tend to be short. Nonetheless, if a feature appears more than once in a sentence, this means that it is important and the frequency value representation captures this aspect.

The selected features are words (not lemmatized) delimited by spaces and simple punctuation marks: *space, (,) , [,] , . , ' , _* that appeared at least three times in the training collection and contain at least an alpha-numeric character, are not part of an English list of stop words[5] and are longer than three characters. Stop words are function words that appear in every document (e.g., *the, it, of, an*) and therefore do not help in classification. The frequency threshold of three is commonly used for text collections because it removes non-informative features and also strings of characters that might be the result of a wrong tokenization when splitting the text into words. Words that have length of one or two characters are not considered as features because of two reasons: possible incorrect tokenization and problems with very short acronyms in the medical domain that could be highly ambiguous (could be a medical acronym or an abbreviation of a common word).

4.2 NLP and biomedical concepts representation

The second type of representation is based on NLP information – noun-phrases, verb-phrases and biomedical concepts (Biomed). In order to extract this type of information from the data, we used the Genia[6] tagger. The tagger analyzes English sentences and outputs the base forms, part-of-speech tags, chunk tags, and named entity tags. The tagger is specifically tuned for biomedical text such as Medline abstracts.

Figure 1 presents an output example by the Genia tagger for the sentence: "*Inhibition of NF-kappaB activation reversed the anti-apoptotic effect of isochamaejasmin.*". The tag O stands for Outside, B for Beginning, and I for Inside.

Figure 1. Example of Genia tagger output

Inhibition	Inhibition	NN	B-NP	O
of	of	IN	B-PP	O
NF-kappaB	NF-kappaB	NN	B-NP	B-protein
activation	activation	NN	I-NP	O
reversed	reverse	VBD	B-VP	O
the	the	DT	B-NP	O
anti-apoptotic	anti-apoptotic	JJ	I-NP	O
effect	effect	NN	I-NP	O
of	of	IN	B-PP	O
isochamaejasmin	isochamaejasmin	NN	B-NP	O
.	.	.	O	O

The noun-phrases and verb-phrases identified by the tagger are considered as features for our second representation technique. The following preprocessing steps are applied before defining the set of final features: remove features that contain only punctuation, remove stop-words, and consider valid features only the lemma-based forms of the identified noun-phrases, verb-phrases and biomedical concepts. The reason to do this is because there are a lot of inflected forms (*e.g.,* plural forms) for the same word and the lemmatized form (the base form of a word) will give us the same base form for all the inflected forms.

4.3 Medical concepts (UMLS) representation

In order to work with a representation that provides features that are more general than the words in the abstracts (used in the BOW representation), we also used the unified medical language system[7] (here on UMLS) concept representations. UMLS is a knowledge source developed at the U.S. National Library of Medicine (here on NLM) and it contains a meta-thesaurus, a semantic network, and the specialist lexicon for biomedical domain. The meta-thesaurus is organized around concepts and meanings; it links alternative names and views of the same concept and identifies useful relationships between different concepts. UMLS contains over 1 million medical concepts, and over 5 million concept names which are hierarchical organized. Each unique concept that is present in the thesaurus has associated multiple text strings variants (slight morphological variations of the concept). All concepts are assigned at least one semantic type from the semantic network providing a generalization of the existing relations between concepts. There are 135 semantic types in the knowledge base linked through 54 relationships.

[5] http://www.site.uottawa.ca/~diana/csi5180/StopWords
[6] http://www-tsujii.is.s.u-tokyo.ac.jp/GENIA/tagger/

[7] http://www.nlm.nih.gov/pubs/factsheets/umls.html

In addition to the UMLS knowledge base, NLM created a set of tools that allow easier access to the useful information. MetaMap[8] is a tool created by NLM that maps free text to medical concepts in the UMLS, or equivalently, it discovers meta-thesaurus concepts in text. With this software, text is processed through a series of modules that in the end will give a ranked list of all possible concept candidates for a particular noun-phrase. For each of the noun phrases that the system finds in the text, variant noun phrases are generated. For each of the variant noun phrases, candidate concepts (concepts that contain the noun phrase variant) from the UMLS meta-thesaurus are retrieved and evaluated. The retrieved concepts are compared to the actual phrase using a fit function that measures the text overlap between the actual phrase and the candidate concept (it returns a numerical value). The best of the candidates are then organized according to the decreasing value of the fit function. We used the top concept candidate for each identified phrase in an abstract as a feature. Figure 2 presents an example of the output of the MetaMap system for the phrase *"to an increased risk"*. The information presented in the brackets, the semantic type, "Qualitative Concept, Quantitative Concept" for the candidate with the fit function value 861 is the feature used for our UMLS representation.

Figure 2. Example of MetaMap system output

Meta Candidates (6)
861 Risk [Qualitative Concept, Quantitative Concept]
694 Increased (Increased (qualifier value)) [Functional Concept]
623 Increase (Increase (qualifier value)) [Functional Concept]
601 Acquired (Acquired (qualifier value)) [Temporal Concept]
601 Obtained (Obtained (attribute)) [Functional Concept]
588 Increasing (Increasing (qualifier value)) [Functional Concept]

Another reason to use a UMLS concept representation is the ***concept drift*** phenomenon that can appear in a BOW representation. Especially in the medical domain texts, this is a frequent problem as stated by Cohen *et al.* (2004). New articles that publish new research on a certain topic bring with them new terms that might not match the ones that were seen in the training process in a certain moment of time.

[8] http://mmtx.nlm.nih.gov/

Experiments for the task tackled in our research are performed with all the above-mentioned representations, plus combinations of them. We combine the BOW, UMLS and NLP and biomedical concepts by putting all features together to represent an instance.

5 Results

This section presents the results obtained for the task of identifying semantic relations with the methods described above. As evaluation measures, we report F-measure and accuracy values. The main evaluation metric that we consider is the F-measure[9], since it is a suitable when the data set is imbalanced. We report the accuracy measure as well, because we want to compare our results with previous work. Table A1 from appendix A presents the results that we obtained with our methods. The table contains F-measure scores for all three semantic relations with the three experimental settings proposed for all combinations of representation and classification algorithms. In this section, since we cannot report all the results for all the classification algorithms, we decided to report the classifiers that obtained the lower and upper margin of results for every representation setting. More detailed descriptions for the results are present in appendix A. We consider as baseline a classifier that always predicts the majority class. For the relation *Cure* the F-measure baseline is 66.51%, for *Prevent* and *Side Effect* 0%.

The next three figures present the best results obtained for the three experimental settings.

Figure 3. Best results for Setting 1.

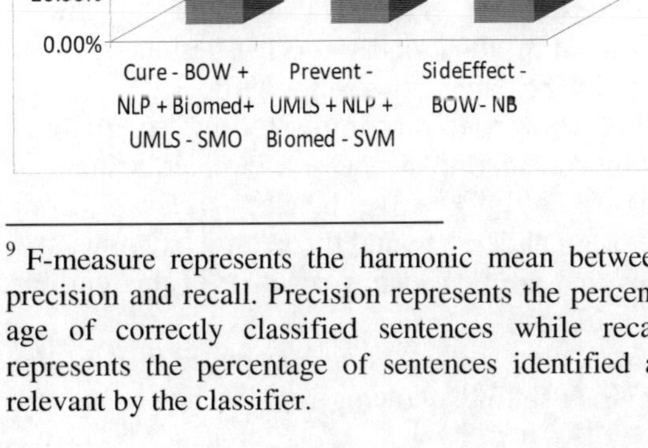

[9] F-measure represents the harmonic mean between precision and recall. Precision represents the percentage of correctly classified sentences while recall represents the percentage of sentences identified as relevant by the classifier.

Figure 4. Best results for Setting 2.

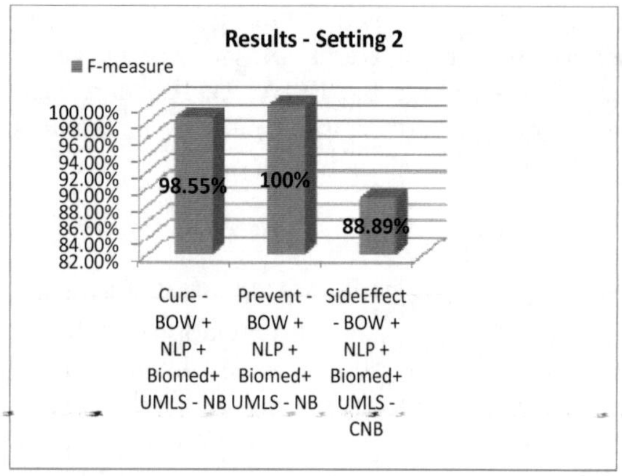

Figure 5. Best results for Setting 3.

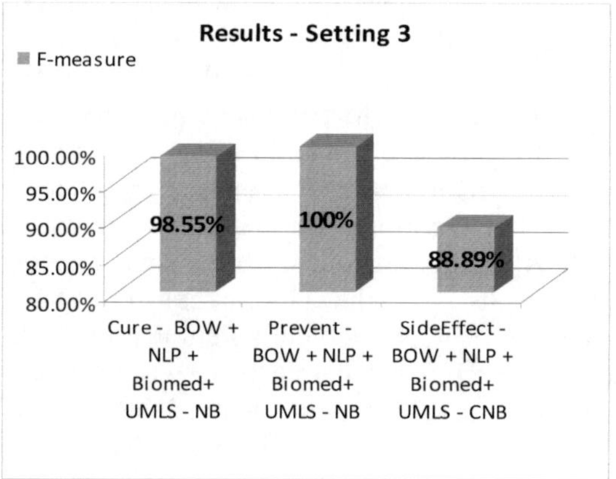

6 Discussion

Our goal was to obtain high performance results for the three semantic relations. The first set of experiments was influenced by previous work on a different task. The results obtained show that this setting might not be suitable for the medical domain, due to one of the following possible explanations: the number of examples that are considered as being neutral is not sufficient or not appropriate (the neutral examples are considered sentences that are annotated with a *Vague* relation in the original data); or the negative examples are not appropriate (the negative examples are considered sentences that talk about either treatment or about diseases). The results of these experiments are shown in Figure 3. As future work, we want to run similar setting experiments when considering negative examples sentences that are not informative, labeled *Irrelevant*, from the original data set, and the neutral examples the ones that are considered negative in this current experiments.

In Setting 2, the results are better than in the previous setting, showing that the neutral examples used in the previous experiments confused the algorithms and were not appropriate. These results validate the fact that the previous setting was not the best one for the task.

The best results for the task are obtained with the third setting, when a model is built and trained on a data set that contains all sentences annotated with the three relations. The representation and the classification algorithms were able to make the distinction between the relations and obtained the best results for this task. The results are: 98.55% F-measure for the *Cure* class, 100% F-measure for the *Prevent* class, and 88.89% for the *Side Effect* class.

Some important observations can be drawn from the obtained results: probabilistic and linear models combined with informative feature representations bring the best results. They are consistent in outperforming the other classifiers in all the three settings. AdaBoost classifier was outperformed by other classifiers, which is a little surprising, taking into consideration the fact that this classifier tends to work better on imbalanced data. BOW is a representation technique that even though it is simplistic, most of the times it is really hard to outperform. One of the major contributions of this work is the fact that the current experiments show that additional information used in the representation settings brings improvements for the task. The task itself is a knowledge-charged task and the experiments show that classifiers can perform better when richer information (*e.g.* concepts for medical ontologies) is provided.

6.1 Comparison to previous work

Even though our main focus is on the three relations mentioned earlier, in order to validate our methodology, we also performed the 8-class classification task, similar to the one done by Rosario and Hearst (2004). Figure 3 presents a graphical comparison of the results of our methods to the ones obtained in the previous work. We report accuracy values for these experiments, as it was done in the previous work.

In Figure 3, the first set of bar-results represents the best individual results for each relation. The representation technique and classification model that obtains the best results are the ones described on the x-axis.

Figure 3. Comparison of results.

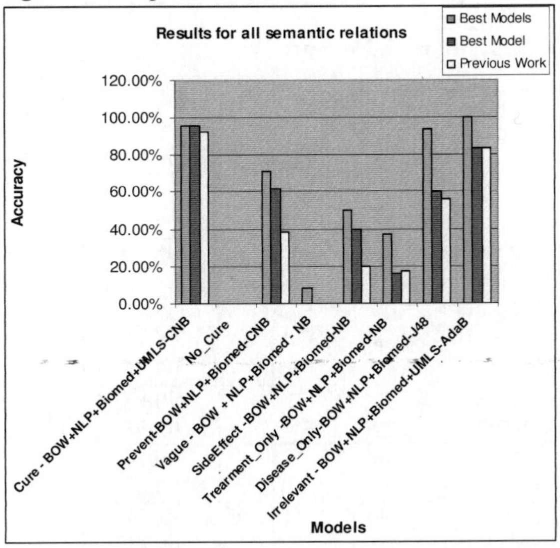

The second series of results represents the overall best model that is reported for each relation. The model reported here is a combination of BOW, verb and noun-phrases, biomedical and UMLS concepts, with a CNB classifier.

The third series of results represent the accuracy results obtained in previous work by Rosario and Hearst (2004). As we can see from the figure, the best individual models have a major improvement over previous results. When a single model is used for all relations, our results improve the previous ones in four relations with the difference varying from: 3 percentage point difference (*Cure*) to 23 percentage point difference (*Prevent*). We obtain the same results for two semantic relations, *No_Cure* and *Vague* and we believe that this is the case due to the fact that these two classes are significantly under-represented compared to the other ones involved in the task. For the *Treatment_Only* relation our results are outperformed with 1.5 percentage points and for the *Irrelevant* relation with 0.1 percentage point, only when we use the same model for all relations.

7 Conclusion and Future Work

We can conclude that additional knowledge and deeper analysis of the task and data in question are required in order to obtain reliable results. Probabilistic models are stable and reliable for the classification of short texts in the medical domain. The representation techniques highly influence the results, common for the ML community, but more informative representations where the ones that consistently obtained the best results.

As future work, we would like to extend the experimental methodology when the first setting is applied, and to use additional sources of information as representation techniques.

References

Ahlers C., Fiszman M., Fushman D., Lang F.-M., Rindflesch T. 2007. *Extracting semantic predications from Medline citations for pharmacogenomics.* Pacific Symposium on Biocomputing, 12:209-220.

Craven M. 1999. *Learning to extract relations from Medline.* AAAI-99 Workshop on Machine Learning for Information Extraction.

Feldman R. Regev Y., Finkelstein-Landau M., Hurvitz E., and Kogan B. 2002. *Mining biomedical literature using information extraction.* Current Drug Discovery.

Friedman C., Kra P., Yu H., Krauthammer M., and Rzhetzky A. 2001. *Genies: a natural-language processing system for the extraction of molecular pathways from journal articles.* Bioinformatics, 17(1).

Ginsberg J., Mohebbi Matthew H., Rajan S. Patel, Lynnette Brammer, Mark S. Smolinski & Larry Brilliant. 2009. *Detecting influenza epidemics using search engine query data.* Nature 457, 1012-1014.

Hunter Lawrence and K. Bretonnel Cohen. 2006. *Biomedical Language Processing: What's Beyond PubMed?* Molecular Cell 21, 589–594.

Ray S. and Craven M. 2001. *Representing sentence structure in Hidden Markov Models for information extraction.* Proceedings of IJCAI-2001.

Rosario B. and Marti A. Hearst. 2004. *Classifying semantic relations in bioscience text.* Proceedings of the 42nd Annual Meeting on Association for Computational Linguistics, 430.

Koppel M. and J. Schler. 2005. *Using Neutral Examples for Learning Polarity,* Proceedings of IJCAI, Edinburgh, Scotland.

Srinivasan P. and T. Rindflesch 2002. *Exploring text mining from Medline.* Proceedings of the AMIA Symposium.

Appendix A. Detailed Results.

Relation	Representation	Setting1		Setting2		Setting3	
		Classification Algorithm - F-Measure (%)					
Cure	NLP+Biomed	AdaB	32.22	AdaB	35.69	CNB	87.88
		ZeroR	66.51	ZeroR	67.48	SVM	94.85
	BOW	AdaB	63.60	AdaB	67.23	CNB	92.57
		CNB	79.22	SVM	81.43	NB	96.80
	UMLS	AdaB	61.08	AdaB	64.78	CNB	88.20
		NB	74.73	NB	76.04	SVM	95.62
	BOW+UMLS	AdaB	56.07	AdaB	74.68	J48	96.13
		CNB	84.54	NB	86.48	NB	97.50
	NLP+Biomed +UMLS	AdaB	61.08	AdaB	64.78	CNB	90.87
		NB	75.18	NB	76.70	SVM	96.58
	NLP+Biomed +BOW	AdaB	53.04	AdaB	77.46	J48	96.14
		SVM	78.98	CNB	81.86	NB	97.86
	NLP+Biomed+ BOW+UMLS	AdaB	53.04	AdaB	72.32	J48	96.32
		SVM	**85.14**	**SVM**	**87.10**	**NB**	**98.55**
Prevent	NLP+Biomed	AdaB	0	AdaB,J48	0	Ada,J48	0
		NB	17.02	NB	22.86	CNB	55.17
	BOW	CNB	31.78	J48	0	SVM	50
		NB	50	NB	61.9	CNB	89.47
	UMLS	AdaB	0	J48	0	J48	0
		NB	28.57	SVM	48.28	CNB	68.75
	BOW+UMLS	J48	39.02	J48	9.09	AdaB	60
		NB	57.14	NB	75.68	CNB	89.47
	NLP+Biomed +UMLS	AdaB	0	J48	16	J48	0
		SVM	**62.50**	SVM	57.69	CNB	97.56
	NLP+Biomed +BOW	SVM	35	J48	0	AdaB	64.52
		NB	54.90	NB	66.67	CNB	92.31
	NLP+Biomed+ BOW+UMLS	J48	30.77	J48	0	AdaB,J48	64.52
		NB	62.30	**SVM**	**77.78**	**NB**	**100**
Side Effect	NLP+Biomed	AdaB	0	J48,SVM	0	AdaB,J48	0
		NB,CNB	7.69	AdaB	18.18	CNB	33.33
	BOW	AdaB	0	AdaB,J48	0	Ada,J48	0
		NB	**34.48**	NB	50	CNB	66.67
	UMLS	AdaB,J48,	0	J48,SVM	0	AdaB,J48	0
		SVM NB	22.22	NB	33.33	NB,CNB	46.15
	BOW+UMLS	AdaB,J48	0	J48	0	AdaB	0
		NB	21.43	NB	47	CNB	75
	NLP+Biomed+ UMLS	AdaB,J48	0	J48	0	AdaB.J48	0
		NB	19.35	NB	31.58	NB,CNB	46.15
	NLP+Biomed+ BOW	AdaB,J48	0	J48	0	AdaB,J48	0
		NB	33.33	**NB**	**55.56**	**CNB**	**88.89**
	NLP+Biomed+ BOW+UMLS	AdaB,J48	0	J48	0	AdaB	0
		NB	24	NB	46.15	**CNB**	**88.89**

Table A1. Results obtained with our methods.

The *Representation* column describes all the feature representation techniques that we tried. The acronym *NLP* stands from verb and noun-phrase features put together and *Biomed* for bio-medical concepts (the ones extracted by Genia tagger). The first line of results for every representation technique presents the classier that obtained the lowest results, while the second line represents the classifier with the best F-measure score. In bold we mark the best scores for all semantic relations in each of the three settings.

Identifying the Information Structure of Scientific Abstracts: An Investigation of Three Different Schemes

Yufan Guo
University of Cambridge, UK
yg244@cam.ac.uk

Anna Korhonen
University of Cambridge, UK
alk23@cam.ac.uk

Maria Liakata
Aberystwyth University, UK
mal@aber.ac.uk

Ilona Silins
Karolinska Institutet, SWEDEN
Ilona.Silins@ki.se

Lin Sun
University of Cambridge, UK
ls418@cam.ac.uk

Ulla Stenius
Karolinska Institutet, SWEDEN
Ulla.Stenius@ki.se

Abstract

Many practical tasks require accessing specific types of information in scientific literature; e.g. information about the objective, methods, results or conclusions of the study in question. Several schemes have been developed to characterize such information in full journal papers. Yet many tasks focus on abstracts instead. We take three schemes of different type and granularity (those based on section names, argumentative zones and conceptual structure of documents) and investigate their applicability to biomedical abstracts. We show that even for the finest-grained of these schemes, the majority of categories appear in abstracts and can be identified relatively reliably using machine learning. We discuss the impact of our results and the need for subsequent task-based evaluation of the schemes.

1 Introduction

Scientific abstracts tend to be very similar in terms of their information structure. For example, many abstracts provide some background information before defining the precise objective of the study, and the conclusions are typically preceded by the description of the results obtained.

Many readers of scientific abstracts are interested in specific types of information only, e.g. the general background of the study, the methods used in the study, or the results obtained. Accordingly, many text mining tasks focus on the extraction of information from certain parts of abstracts only. Therefore classification of abstracts (or full articles) according to the categories of information structure can support both the manual study of scientific literature as well as its automatic analysis, e.g. information extraction, summarization and information retrieval (Teufel and Moens, 2002; Mizuta et al., 2005; Tbahriti et al., 2006; Ruch et al., 2007).

To date, a number of different schemes and techniques have been proposed for sentence-based classification of scientific literature according to information structure, e.g. (Teufel and Moens, 2002; Mizuta et al., 2005; Lin et al., 2006; Hirohata et al., 2008; Teufel et al., 2009; Shatkay et al., 2008; Liakata et al., 2010). Some of the schemes are coarse-grained and merely classify sentences according to typical section names seen in scientific documents (Lin et al., 2006; Hirohata et al., 2008). Others are finer-grained and based e.g. on argumentative zones (Teufel and Moens, 2002; Mizuta et al., 2005; Teufel et al., 2009), qualitative dimensions (Shatkay et al., 2008) or conceptual structure (Liakata et al., 2010) of documents.

The majority of such schemes have been developed for full scientific journal articles which are richer in information and also considered to be more in need of the definition of information structure (Lin, 2009). However, many practical tasks currently focus on abstracts. As a distilled summary of key information in full articles, abstracts may exhibit an entirely different distribution of scheme categories than full articles. For tasks involving abstracts, it would be useful to know which schemes are applicable to abstracts and which can be automatically identified in them with reasonable accuracy.

In this paper, we will compare the applicability of three different schemes – those based on section names, argumentative zones and conceptual structure of documents – to a collection of biomedical abstracts used for cancer risk assessment (CRA). CRA is an example of a real-world task which could greatly benefit from knowledge about the information structure of abstracts since cancer risk assessors look for a variety of information in them ranging from specific methods to

Proceedings of the 2010 Workshop on Biomedical Natural Language Processing, ACL 2010, pages 99–107,
Uppsala, Sweden, 15 July 2010.

results concerning different chemicals (Korhonen et al., 2009). We report work on the annotation of CRA abstracts according to each scheme and investigate the schemes in terms of their distribution, mutual overlap, and the success of identifying them automatically using machine learning. Our investigation provides an initial idea of the practical usefulness of the schemes for tasks involving abstracts. We discuss the impact of our results and the further task-based evaluation which we intend to conduct in the context of CRA.

2 The three schemes

We investigate three different schemes – those based on Section Names (S1), Argumentative Zones (S2) and Core Scientific Concepts (S3):

S1: The first scheme differs from the others in the sense that it is actually designed for abstracts. It is based on section names found in some scientific abstracts. We use the 4-way classification from (Hirohata et al., 2008) where abstracts are divided into objective, method, results and conclusions. Table 1 provides a short description of each category for this and other schemes (see also this table for any category abbreviations used in this paper).

S2: The second scheme is based on Argumentative Zoning (AZ) of documents. The idea of AZ is to follow the knowledge claims made by authors. Teufel and Moens (2002) introduced AZ and applied it to computational linguistics papers. Mizuta et al. (2005) modified the scheme for biology papers. More recently, Teufel et al. (2009) introduced a refined version of AZ and applied it to chemistry papers. As these schemes are too fine-grained for abstracts (some of the categories do not appear in abstracts at all), we adopt a reduced version of AZ which integrates seven categories from (Teufel and Moens, 2002) and (Mizuta et al., 2005) - those which actually appear in abstracts.

S3: The third scheme is concept-driven and ontology-motivated (Liakata et al., 2010). It treats scientific papers as humanly-readable representations of scientific investigations and seeks to retrieve the structure of the investigation from the paper as generic high-level Core Scientific Concepts (CoreSC). The CoreSC is a 3-layer annotation scheme but we only consider the first layer in the current work. The second layer pertains to properties of the categories (e.g. "advantage" vs. "disadvantage" of METH, "new" vs. "old" METH or OBJT). Such level of granularity is rare in abstracts. The 3rd layer involves coreference identification between the same instances of each category, which is also not of concern in abstracts. With eleven categories, S3 is the most fine-grained of our schemes. CoreSC has been previously applied to chemistry papers (Liakata et al., 2010, 2009).

3 Data: cancer risk assessment abstracts

We used as our data the corpus of CRA abstracts described in (Korhonen et al., 2009) which contains MedLine abstracts from different subdomains of biomedicine. The abstracts were selected so that they provide rich information about various scientific data (human, animal and cellular) used for CRA. We selected 1000 abstracts (in random) from this corpus. The resulting data includes 7,985 sentences and 225,785 words in total.

4 Annotation of abstracts

Annotation guidelines. We used the guidelines of Liakata for S3 (Liakata and Soldatova, 2008), and developed the guidelines for S1 and S2 (15 pages each). The guidelines define the unit (a sentence) and the categories of annotation and provide advice for conflict resolution (e.g. which categories to prefer when two or several are possible within the same sentence), as well as examples of annotated abstracts.

Annotation tool. We modified the annotation tool of Korhonen et al. (2009) so that it could be used to annotate abstracts according to the schemes. This tool was originally developed for the annotation of CRA abstracts according to the scientific evidence they contain. The tool works as a Firefox plug-in. Figure 1 shows an example of an abstract annotated according to the three schemes.

Description of annotation. Using the guidelines and the tool, the CRA corpus was annotated according to each of the schemes. The annotation proceeded scheme by scheme, independently, so that annotations of one scheme were not based on any of the other two. One annotator (a computational linguist) annotated all the abstracts according to the three schemes, starting from the coarse-grained S1, then proceeding to S2 and finally to the finest-grained S3. It took 45, 50 and 90 hours in total for S1, S2 and S3, respectively.

The resulting corpus. Table 2 shows the distribution of sentences per scheme category in the resulting corpus.

Table 1: The Three Schemes

S1	Objective	OBJ	The background and the aim of the research
	Method	METH	The way to achieve the goal
	Result	RES	The principle findings
	Conclusion	CON	Analysis, discussion and the main conclusions
S2	Background	BKG	The circumstances pertaining to the current work, situation, or its causes, history, etc.
	Objective	OBJ	A thing aimed at or sought, a target or goal
	Method	METH	A way of doing research, esp. according to a defined and regular plan; a special form of procedure or characteristic set of procedures employed in a field of study as a mode of investigation and inquiry
	Result	RES	The effect, consequence, issue or outcome of an experiment; the quantity, formula, etc. obtained by calculation
	Conclusion	CON	A judgment or statement arrived at by any reasoning process; an inference, deduction, induction; a proposition deduced by reasoning from other propositions; the result of a discussion, or examination of a question, final determination, decision, resolution, final arrangement or agreement
	Related work	REL	A comparison between the current work and the related work
	Future work	FUT	The work that needs to be done in the future
S3	Hypothesis	HYP	A statement that has not been yet confirmed rather than a factual statement
	Motivation	MOT	The reason for carrying out the investigation
	Background	BKG	Description of generally accepted background knowledge and previous work
	Goal	GOAL	The target state of the investigation where intended discoveries are made
	Object	OBJT	An entity which is a product or main theme of the investigation
	Experiment	EXP	Experiment details
	Model	MOD	A statement about a theoretical model or framework
	Method	METH	The means by which the authors seek to achieve a goal of the investigation
	Observation	OBS	The data/phenomena recorded within an investigation
	Result	RES	Factual statements about the outputs of an investigation
	Conclusion	CON	Statements inferred from observations and results, relating to research hypothesis

Inter-annotator agreement. We measured the inter-annotator agreement on 300 abstracts (i.e. a third of the corpus) using three annotators (one linguist, one expert in CRA, and the computational linguist who annotated all the corpus). According to Cohen's Kappa (Cohen, 1960), the inter-annotator agreement for S1, S2, and S3 was $\kappa = 0.84$, $\kappa = 0.85$, and $\kappa = 0.50$, respectively. According to (Landis and Koch, 1977), the agreement 0.81-1.00 is perfect and 0.41-0.60 is moderate. Our results indicate that S1 and S2 are the easiest schemes for the annotators and S3 the most challenging. This is not surprising as S3 is the scheme with the finest granularity. Its reliable identification may require a longer period of training and possibly improved guidelines. Moreover, previous annotation efforts using S3 have used domain experts for annotation (Liakata et al., 2009, 2010). In our case the domain expert and the linguist agreed the most on S3 ($\kappa = 0.60$). For S1 and S2 the best agreement was between the linguist and the computational linguist ($\kappa = 0.87$ and $\kappa = 0.88$, respectively).

Table 2: Distribution of sentences in the scheme-annotated CRA corpus

S1	OBJ	METH	RES	CON								
	61483	39163	89575	35564	Words							
	2145	1396	3203	1241	Sentences							
	27%	17%	40%	16%	Sentences							
S2	BKG	OBJ	METH	RES	CON	REL	FUT					
	36828	23493	41544	89538	30752	2456	1174	Words				
	1429	674	1473	3185	1082	95	47	Sentences				
	18%	8%	18%	40%	14%	1%	1%	Sentences				
S3	HYP	MOT	BKG	GOAL	OBJT	EXP	MOD	METH	OBS	RES	CON	
	2676	4277	28028	10612	15894	22444	1157	17982	17402	75951	29362	Words
	99	172	1088	294	474	805	41	637	744	2582	1049	Sentences
	1%	2%	14%	4%	6%	10%	1%	8%	9%	32%	13%	Sentences

5 Comparison of the schemes in terms of annotations

The three schemes we have used to annotate abstracts were developed independently and have separate guidelines. Thus, even though they seem to have some categories in common (e.g. METH, RES, CON) this does not necessarily guarantee that the latter cover the same information across all three schemes. We therefore wanted to investigate the relation between the schemes and the extent of overlap or complementarity between them.

We used the annotations obtained with each scheme to create three contingency matrices for pairwise comparison. We calculated the chi-squared Pearson statistic, the chi-squared like-

Figure 1: An example of an abstract annotated according to the three schemes

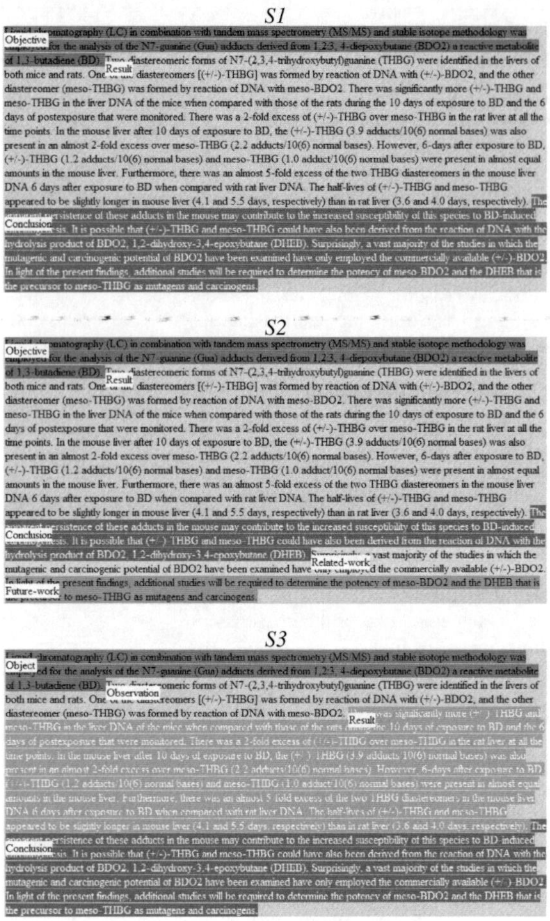

lihood ratio, the contingency coefficient and Cramer's V (Table 3)[1], all of which showed a definite correlation between rows and columns for the pairwise comparison of all three schemes.

However, none of the above measures give an indication of the differential association between schemes, i.e. whether it goes both directions and to what extent. For this reason we calculated the Goodman-Kruskal lambda L statistic (Siegel and Castellan, 1988), which gives us the reduction in error for predicting the categories of one annotation scheme, if we know the categories assigned according to the other. When using the categories of S1 as the independent variables, we obtained a lambda of over 0.72 which suggests a 72% reduction in error in predicting S2 categories and 47% in predicting S3 categories. With S2 categories being the independent variables, we obtained a reduction in error of 88% when predicting S1 and 55% when predicting S3 categories. The lower lambdas for predicting S3 are hardly surprising as S3 has 11 categories as opposed to 4 and 7 for S1 and S2 respectively. S3 on the other hand has strong predictive power in predicting the categories of S1 and S2 with lambdas of 0.86 and 0.84 respectively. In terms of association, S1 and S2 seem to be more strongly associated, followed by S1 and S3 and then S2 and S3.

We were then interested in the correspondence between the actual categories of the three schemes, which is visualized in Figure 2. Looking at the categories of S1, OBJ maps mostly to BKG and OBJ in S2 (with a small percentage in METH and REL). S1 OBJ maps to BKG, GOAL, HYP, MOT and OBJT in S3 (with a small percentage in METH and MOD). S1 METH maps to METH in S2 (with a small percentage in S2 OBJ) while it maps to EXP, METH and MOD in S3 (with a small percentage in GOAL and OBJT). S1 RES covers S2 RES and 40% REL, whereas in S3 it covers RES, OBS and 20% MOD. S1 CON covers S2 CON, FUT, 45% REL and a small percentage of RES. In terms of the S2 vs S3 comparison, S2 BKG maps to S3 BKG, HYP, MOT and a small percentage of OBJT and MOD. S2 CON maps to S3 CON, with a small percentage in RES, OBS and HYP. S2 FUT maps entirely to S3 CON. S2 METH maps to S3 METH, EXP, MOD, 20% OBJT and a small percentage of GOAL. S2 OBJ maps to S3 GOAL and OBJT, with 15% HYP, MOD and MOT and a small percentage in METH. S2 REL spans across S3 CON, RES, MOT and OBJT, albeit in very small percentages. Finally, S2 RES maps to S3 RES and OBS, with 25% in MOD and small percentages in METH, CON, OBJT. Thus, it appears that each category in S1 maps to a couple of categories in S2 and several in S3, which in turn seem to elaborate on the S2 categories.

Based on the above analysis of the categories, it is reasonable to assume a subsumption relation between the categories of the type S1 $>$ S2 $>$ S3, with REL cutting across several of the S3 categories and FUT branching off S3 CON. This is an interesting and exciting outcome given that the three different schemes have such a different origin.

[1]These are association measures for r x c tables. We used the implementation in the vcd package of R (http://www.r-project.org/).

Table 3: Association measures between schemes S1, S2, S3

	S1 vs S2			S1 vs S3			S2 vs S3		
	X^2	**df**	P	X^2	**df**	P	X^2	**df**	P
Likelihood Ratio	5577.1	18	0	5363.6	30	0	6293.4	60	0
Pearson	6613.0	18	0	6371.0	30	0	8554.7	60	0
Contingency Coeff	0.842			0.837			0.871		
Cramer's V	0.901			0.885			0.725		

Figure 2: Pairwise interpretation of categories of one scheme in terms of the categories of the other.

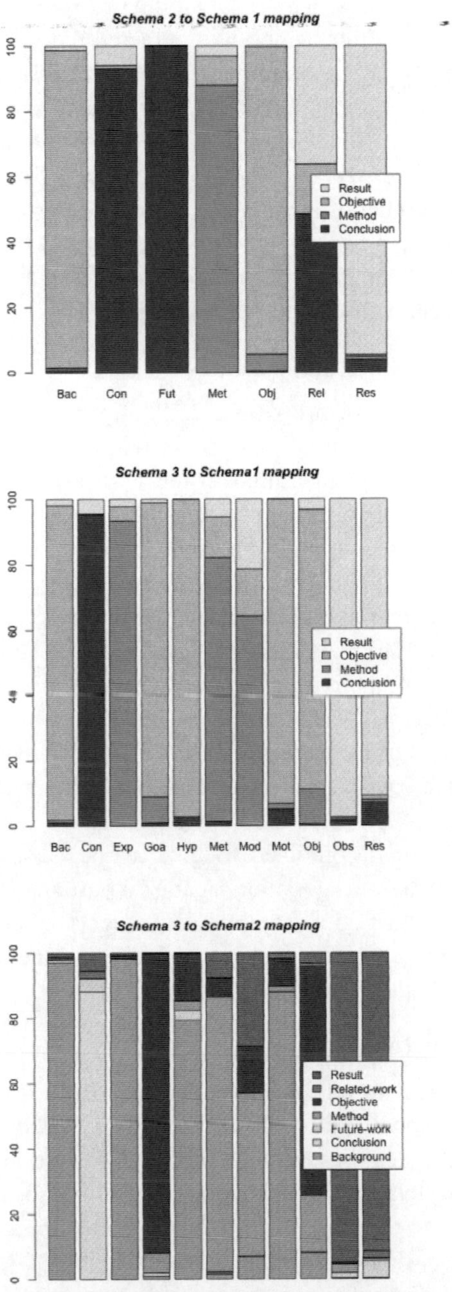

6 Automatic identification of information structure

6.1 Features

The first step in automatic identification of information structure is feature extraction. We chose a number of general purpose features suitable for all the three schemes. With the exception of our novel verb class feature, the features are similar to those employed in related works, e.g. (Teufel and Moens, 2002; Mullen et al., 2005; Hirohata et al., 2008):

History. There are typical patterns in the information structure, e.g. RES tends to be followed by CON rather than by BKG. Therefore, we used the category assigned to the previous sentence as a feature.

Location. Categories tend to appear in typical positions in a document, e.g. BKG occurs often in the beginning and CON at the end of the abstract. We divided each abstract into ten equal parts (1-10), measured by the number of words, and defined the location (of a sentence) feature by the parts where the sentence begins and ends.

Word. Like many text classification tasks, we employed all the words in the corpus as features.

Bi-gram. We considered each bi-gram (combination of two word features) as a feature.

Verb. Verbs are central to the meaning of sentences, and can vary from one category to another. For example, *experiment* is frequent in METH and *conclude* in CON. Previous works have used the matrix verb of each sentence as a feature. Because the matrix verb is not the only meaningful verb, we used all the verbs instead.

Verb Class. Because individual verbs can result in sparse data problems, we also experimented with a novel feature: verb class (e.g. the class of EXPERIMENT verbs for verbs such as *measure* and *inject*). We obtained 60 classes by clustering verbs appearing in full cancer risk assessment articles using the approach of Sun and Korhonen (2009).

POS. Tense tends to vary from one category to another, e.g. past is common in RES and past participle in CON. We used the part-of-speech (POS) tag of each verb assigned by the C&C tagger (Curran et al., 2007) as a feature.

GR. Structural information about heads and dependents has proved useful in text classification. We used grammatical relations (GRs) returned by the C&C parser as features. They consist of a named relation, a head and a dependent, and possibly extra parameters depending on the relation involved, e.g. *(dobj investigate mouse)*. We created features for each subject (*ncsubj*), direct object (*dobj*), indirect object (*iobj*) and second object (*obj2*) relation in the corpus.

Subj and Obj. As some GR features may suffer from data sparsity, we collected all the subjects and objects (appearing with any verbs) from GRs and used them as features.

Voice. There may be a correspondence between the active and passive voice and categories (e.g. passive is frequent in METH). We therefore used voice as a feature.

6.2 Methods

We used Naive Bayes (NB) and Support Vector Machines (SVM) for classification. NB is a simple and fast method while SVM has yielded high performance in many text classification tasks.

NB applies Bayes' rule and Maximum Likelihood estimation with strong independence assumptions. It aims to select the class c with maximum probability given the feature set F:

$$\arg\max_c P(c|F) = \arg\max_c \frac{P(c)\cdot P(F|c)}{P(F)}$$
$$= \arg\max_c P(c)\cdot P(F|c)$$
$$= \arg\max_c P(c)\cdot \prod_{f\in F} P(f|c)$$

SVM constructs hyperplanes in a multidimensional space that separates data points of different classes. Good separation is achieved by the hyperplane that has the largest distance from the nearest data points of any class. The hyperplane has the form $w \cdot x - b = 0$, where w is the normal vector to the hyperplane. We want to maximize the distance from the hyperplane to the data points, or the distance between two parallel hyperplanes each of which separates the data. The parallel hyperplanes can be written as:

$w\cdot x - b = 1$ and $w\cdot x - b = -1$, and the distance between the two is $\frac{2}{|w|}$. The problem reduces to:

Minimize $|w|$

Subject to $w \cdot x_i - b \geq 1$ for x_i of one class, and $w \cdot x_i - b \leq -1$ for x_i of the other.

7 Experimental evaluation

7.1 Preprocessing

We developed a tokenizer to detect the boundaries of sentences and to perform basic tokenisation, such as separating punctuation from adjacent words e.g. in tricky biomedical terms such as *2-amino-3,8-diethylimidazo[4,5-f]quinoxaline*. We used the C&C tools (Curran et al., 2007) for POS tagging, lemmatization and parsing. The lemma output was used for extracting *Word*, *Bi-gram* and *Verb* features. The parser produced GRs for each sentence from which we extracted the *GR*, *Subj*, *Obj* and *Voice* features. We only considered the GRs relating to verbs. The "obj" marker in a subject relation indicates a verb in passive voice (e.g. *(ncsubj observed_14 difference_5 obj)*). To control the number of features we removed the words and GRs with fewer than 2 occurrences and bi-grams with fewer than 5 occurrences, and lemmatized the lexical items for all the features.

7.2 Evaluation methods

We used Weka (Witten, 2008) for the classification, employing its NB and SVM linear kernel. The results were measured in terms of accuracy (the percentage of correctly classified sentences), precision, recall, and F-Measure. We used 10-fold cross validation to avoid the possible bias introduced by relying on any one particular split of the data. The data were randomly divided into ten parts of approximately the same size. Each individual part was retained as test data and the remaining nine parts were used as training data. The process was repeated ten times with each part used once as the test data. The resulting ten estimates were then combined to give a final score. We compare our classifiers against a baseline method based on random sampling of category labels from training data and their assignment to sentences on the basis of their observed distribution.

7.3 Results

Table 4 shows F-measure results when using each individual feature alone, and Table 5 when using all the features but the individual feature in question. In these two tables, we only report the results for SVM which performed considerably better than NB. Although we have results for most scheme categories, the results for some are missing due to the lack of sufficient training data (see Table 2), or due to a small feature set (e.g. *History* alone).

Table 4: F-Measure results when using each individual feature alone

		a	b	c	d	e	f	g	h	i	j	k
S1	OBJ	.39	.83	.71	.69	.52	.45	.45	.45	.54	.39	-
	METH	-	.47	.81	.74	.63	.49	-	.46	.03	.42	.51
	RES	-	.76	.85	.86	.76	.70	.72	.69	.70	.68	.54
	CON	-	.72	.70	.65	.63	.53	.49	.57	.68	.20	-
S2	BKG	.26	.73	.69	.67	.45	.38	.56	.33	.33	.29	-
	OBJ	-	.13	.72	.68	.54	.63	-	.49	.48	.20	-
	METH	-	.50	.81	.72	.64	.47	-	.47	.03	.42	.51
	RES	-	.76	.85	.87	.76	.72	.72	.70	.69	.68	.54
	CON	-	.70	.73	.71	.62	.51	.40	.61	.67	.23	-
	REL	-	-	-	-	-	-	-	-	-	-	-
	FUT	-	-	-	-	-	-	-	-	-	-	-
S3	HYP	-	-	-	-	.67	-	-	-	-	-	-
	MOT	.18	.57	.70	.49	.39	.13	.36	.33	.30	.40	-
	BKG	-	-	.54	.40	.21	-	-	.11	.06	.06	-
	GOAL	-	-	.53	.33	.22	-	.19	.31	-	.25	-
	OBJT	-	-	.73	.63	.60	.10	-	.26	.32	-	-
	EXP	-	.22	.63	.46	.33	.30	-	.31	.07	.44	.25
	MOD	-	-	-	-	-	-	-	-	-	-	-
	METH	-	-	.82	.61	.39	.39	-	.50	-	.37	-
	OBS	-	.59	.75	.71	.63	.56	.56	.54	.48	.52	.47
	RES	-	-	.87	.73	.41	.34	-	.38	.24	.35	-
	CON	-	.74	.68	.65	.65	.50	.48	.49	.55	.21	-

a-k: History, Location, Word, Bi-gram, Verb, Verb Class, POS, GR, Subj, Obj, Voice

Table 5: F-Measure results using all the features and all but one of the features

		ALL	A	B	C	D	E	F	G	H	I	J	K
S1	OBJ	.90	.89	.87	.92	.90	.90	.91	.91	.91	.92	.91	.88
	METH	.80	.81	.80	.80	.79	.81	.79	.80	.80	.80	.81	.81
	RES	.88	.90	.88	.90	.88	.90	.88	.88	.88	.89	.89	.90
	CON	.86	.85	.82	.87	.88	.90	.90	.88	.89	.88	.88	.90
S2	BKG	.91	.94	.90	.90	.93	.94	.94	.91	.93	.94	.92	.94
	OBJ	.72	.78	.84	.78	.83	.88	.84	.81	.83	.84	.78	.83
	METH	.81	.83	.80	.81	.80	.85	.80	.78	.81	.81	.82	.83
	RES	.88	.90	.88	.89	.88	.91	.89	.89	.90	.90	.90	.89
	CON	.84	.83	.77	.83	.86	.88	.86	.87	.88	.89	.88	.81
	REL	-	-	-	-	-	-	-	-	-	-	-	-
	FUT	-	1.0	1.0	1.0	1.0	1.0	1.0	1.0	1.0	1.0	1.0	1.0
S3	HYP	-	-	-	-	-	-	-	-	-	-	-	-
	MOT	.82	.84	.80	.76	.82	.82	.83	.78	.83	.83	.83	.83
	BKG	.59	.60	.60	.54	.67	.62	.62	.59	.61	.61	.62	.61
	GOAL	.62	.67	.67	.62	.71	.62	.67	.43	.67	.67	.67	.62
	OBJT	.88	.85	.83	.74	.83	.85	.83	.74	.83	.83	.83	.85
	EXP	.72	.68	.72	.53	.65	.70	.72	.73	.74	.74	.72	.68
	MOD	-	-	-	-	-	-	-	-	-	-	-	-
	METH	.87	.86	.87	.66	.85	.89	.87	.88	.86	.86	.87	.86
	OBS	.82	.81	.84	.72	.80	.82	.81	.80	.82	.82	.81	.81
	RES	.87	.87	.88	.74	.87	.86	.87	.86	.87	.87	.87	.88
	CON	.88	.88	.82	.88	.83	.87	.87	.84	.87	.88	.87	.86

A-K: History, Location, Word, Bi-gram, Verb, Verb Class, POS, GR, Subj, Obj, Voice
We have 1.0 for FUT in S2 probably because the size of the training data is just right, and the model doesn't overfit the data. We make this assumption because we have 1.0 for almost all the categories in the training data, but only for FUT on the test data.

Table 6: Baseline and best NB and SVM results

		Acc.	F-Measure			
S1			OBJ	METH	RES	CON
	BL	.29	.23	.23	.39	.18
	NB	.82	.85	.75	.85	.71
	SVM	.89	.90	.81	.90	.90

		Acc.	F-Measure						
S2			BKG	OBJ	METH	RES	CON	REL	FUT
	BL	.25	.13	.08	.22	.40	.13	-	-
	NB	.76	.79	.25	.70	.83	.66	-	-
	SVM	.90	.94	.88	.85	.91	.88	-	1.0

		Acc.	F-Measure										
S3			HYP	MOT	BKG	GOAL	OBJT	EXP	MOD	METH	OBS	RES	CON
	BL	.15	-	.10	.06	.04	.06	.11	-	.13	.24	.15	.17
	NB	.53	-	.56	-	-	-	.30	-	.32	.61	.59	.62
	SVM	.81	-	.82	.62	.62	.85	.70	-	.89	.82	.86	.87

Looking at individual features alone, *Word*, *Bi-gram* and *Verb* perform the best for all the schemes, and *History* and *Voice* perform the worst. In fact *History* performs very well on the training data, but for the test data we can only use estimates rather than the actual labels. The *Voice* feature works only for RES and METH for S1 and S2, and for OBS for S3. This feature is probably only meaningful for some of the categories.

When using all but one of the features, S1 and S2 suffer the most from the absence of *Location*, while S3 from the absence of *Word/POS*. *Verb Class* on its own performs worse than *Verb*, however when combined with other features it performs better: leave-Verb-out outperforms leave-Verb Class-out.

After comparing the various combinations of features, we found that the best selection of features was *all but the Verb* for all the schemes. Table 6 shows the results for the baseline (BL), and the best results for NB and SVM. NB and SVM perform clearly better than BL for all the schemes. The results for SVM are the best. NB yields the highest performance with S1. Being sensitive to sparse data, it does not perform equally well on S2 and S3 which have a higher number of categories, some of which are low in frequency (see Table 2).

For S1, SVM finds all the four scheme categories with the accuracy of 89%. F-measure is 90 for OBJ, RES and CON and 81 for METH. For S2, the classifier finds six of the seven categories, with the accuracy of 90% and the average F-measure of 91 for the six categories. As with S2, METH has the lowest performance (at 85 F-measure). The one missing category (REL) appears in our abstract data with very low frequency (see Table 2).

For S3, SVM uncovers as many as nine of the 11 categories with accuracy of 81%. Six categories perform well, with F-measure higher than 80. EXP, BKG and GOAL have F-measure of 70, 62 and 62, respectively. Like the missing categories HYP and MOD, GOAL is very low in frequency. The lower performance of the higher frequency EXP and BKG is probably due to low precision in distinguishing between EXP and METH, and BKG and other categories, respectively.

8 Discussion and conclusions

The results from our corpus annotation (see Table 2) show that for the coarse-grained S1, all the four categories appear frequently in biomedical abstracts (this is not surprising because S1 was actually designed for abstracts). All of them can be identified using machine learning. For S2 and S3, the majority of categories appear in abstracts with high enough frequency that we can conclude that also these two schemes are applicable to abstracts. For S2 we identified six categories using machine learning, and for S3 as many as nine, indicating that automatic identification of the schemes in abstracts is realistic.

Our analysis in section 5 showed that there is a subsumption relation between the categories of the schemes. S2 and S3 provide finer-grained information about the information structure of abstracts than S1, even with their 2-3 low frequency (or missing) categories. They can be useful for practical tasks requiring such information. For example, considering S3, there may be tasks where one needs to distinguish between EXP, MOD and METH, between HYP, MOT and GOAL, or between OBS and RES.

Ultimately, the optimal scheme will depend on the level of detail required by the application at hand. Therefore, in the future, we plan to conduct task-based evaluation of the schemes in the context of CRA and to evaluate the usefulness of S1-S3 for tasks cancer risk assessors perform on abstracts (Korhonen et al., 2009). Now that we have annotated the CRA corpus for S1-S3 and have a machine learning approach available, we are in an excellent position to conduct this evaluation.

A key question for real-world tasks is the level of machine learning performance required. We plan to investigate this in the context of our task-based evaluation. Although we employed fairly standard text classification methodology in our experiments, we obtained high performance for S1 and S2. Due to the higher number of categories (and less training data for each of them), the overall performance was not equally impressive for S3 (although still quite high at 81% accuracy).

Hirohata et al. (2008) have showed that the amount of training data can have a big impact on our task. They used c. 50,000 Medline abstracts annotated (by the authors of the Medline abstracts) as training data for S1. When using a small set of standard text classification features and Conditional Random Fields (CRF) (Lafferty et al., 2001) for classification, they obtained 95.5% per-sentence accuracy on 1000 abstracts. However, when only 1000 abstracts were used for training the accuracy was considerably worse; their reported per-abstract accuracy dropped from 68.8% to less than 50%. Although it would be difficult to obtain similarly huge training data for S2 and S3, this result suggests that one key to improved performance is larger training data, and this is what we plan to explore especially for S3.

In addition we plan to improve our method. We showed that our schemes partly overlap and that similar features and methods tend to perform the best / worst for each of the schemes. It is therefore unlikely that considerable scheme specific tuning will be necessary. However, we plan to develop our features further and to make better use of the sequential nature of information structure. Currently this is only represented as the History feature, which provides a narrow window view to the category of the previous sentence. Also we plan to compare SVM against methods such as CRF and Maximum Entropy which have proved successful in recent related works (Hirohata et al., 2008; Merity et al., 2009). The resulting models will be evaluated both directly and in the context of CRA to provide an indication of their practical usefulness for real-world tasks.

Acknowledgments

The work reported in this paper was funded by the Royal Society (UK), the Swedish Research Council, FAS (Sweden), and JISC (UK) which is funding the SAPIENT Automation project. YG was funded by the Cambridge International Scholarship.

References

J. Cohen. 1960. A coefficient of agreement for nominal scales. *Educational and Psychological Measurement*, 20:37–46.

J. R. Curran, S. Clark, and J. Bos. 2007. Linguistically motivated large-scale nlp with c&c and boxer. In *Proceedings of the ACL 2007 Demonstrations Session*, pages 33–36.

K. Hirohata, N. Okazaki, S. Ananiadou, and M. Ishizuka. 2008. Identifying sections in scientific abstracts using conditional random fields. In *Proc. of 3rd International Joint Conference on Natural Language Processing*.

A. Korhonen, L. Sun, I. Silins, and U. Stenius. 2009. The first step in the development of text mining technology for cancer risk assessment: Identifying and organizing scientific evidence in risk assessment literature. *BMC Bioinformatics*, 10:303.

J. Lafferty, A. McCallum, and F. Pereira. 2001. Conditionl random fields: Probabilistic models for segmenting and labeling sequence data. In *Proc. of ICML*.

J. R. Landis and G. G. Koch. 1977. The measurement of observer agreement for categorical data. *Biometrics*, 33:159–174.

M. Liakata and L.N. Soldatova. 2008. Guidelines for the annotation of general scientific concepts. *Aberystwyth University, JISC Project Report http://ie-repository.jisc.ac.uk/88/.*

M. Liakata, Claire Q, and L.N. Soldatova. 2009. Semantic annotation of papers: Interface & enrichment tool (sapient). In *Proceedings of BioNLP-09*, pages 193–200, Boulder, Colorado.

M. Liakata, S. Teufel, A. Siddharthan, and C. Batchelor. 2010. Corpora for the conceptualisation and zoning of scientific papers. *To appear in the 7th International Conference on Language Resources and Evaluation.*

J. Lin, D. Karakos, D. Demner-Fushman, and S. Khudanpur. 2006. Generative content models for structural analysis of medical abstracts. In *Proceedings of BioNLP-06*, pages 65–72, New York, USA.

J. Lin. 2009. Is searching full text more effective than searching abstracts? *BMC Bioinformatics*, 10:46.

S. Merity, T. Murphy, and J. R. Curran. 2009. Accurate argumentative zoning with maximum entropy models. In *Proceedings of the 2009 Workshop on Text and Citation Analysis for Scholarly Digital Libraries*, pages 19–26. Association for Computational Linguistics.

Y. Mizuta, A. Korhonen, T. Mullen, and N. Collier. 2005. Zone analysis in biology articles as a basis for information extraction. *International Journal of Medical Informatics on Natural Language Processing in Biomedicine and Its Applications.*

T. Mullen, Y. Mizuta, and N. Collier. 2005. A baseline feature set for learning rhetorical zones using full articles in the biomedical domain. *Natural language processing and text mining*, 7:52–58.

P. Ruch, C. Boyer, C. Chichester, I. Tbahriti, A. Geissbuhler, P. Fabry, J. Gobeill, V. Pillet, D. Rebholz-Schuhmann, C. Lovis, and A. L. Veuthey. 2007. Using argumentation to extract key sentences from biomedical abstracts. *Int J Med Inform*, 76:195–200.

H. Shatkay, F. Pan, A. Rzhetsky, and W. J. Wilbur. 2008. Multi-dimensional classification of biomedical text: Toward automated, practical provision of high-utility text to diverse users. *Bioinformatics*, 18:2086–2093.

S. Siegel and N. J. Jr. Castellan. 1988. *Nonparametric Statistics for the Behavioral Sciences*. McGraw-Hill, Berkeley, CA, 2nd edition.

L. Sun and A. Korhonen. 2009. Improving verb clustering with automatically acquired selectional preference. In *Proceedings of the Conference on Empirical Methods in Natural Language Processing*.

I. Tbahriti, C. Chichester, Frederique Lisacek, and P. Ruch. 2006. Using argumentation to retrieve articles with similar citations. *Int J Med Inform*, 75:488–495.

S. Teufel and M. Moens. 2002. Summarizing scientific articles: Experiments with relevance and rhetorical status. *Computational Linguistics*, 28:409–445.

S. Teufel, A. Siddharthan, and C. Batchelor. 2009. Towards domain-independent argumentative zoning: Evidence from chemistry and computational linguistics. In *Proc. of EMNLP*.

I. H. Witten, 2008. *Data mining: practical machine learning tools and techniques with Java Implementations*. http://www.cs.waikato.ac.nz/ml/weka/.

Reconstruction of semantic relationships from their projections in biomolecular domain

Juho Heimonen, Jari Björne, and **Tapio Salakoski**
University of Turku
Turku Centre for Computer Science and
Department of Information Technology
Joukahaisenkatu 3–5
20520 Turku, Finland
first.last@utu.fi

Abstract

The extraction of nested, semantically rich relationships of biological entities has recently gained popularity in the biomedical text mining community. To move toward this objective, a method is proposed for reconstructing original semantic relationship graphs from projections, where each node and edge is mapped to the representative of its equivalence class, by determining the relationship argument combinations that represent real relationships. It generalises the limited postprocessing step of the method of Björne et al. (2010) and hence extends this extraction method to arbitrarily deep relationships with unrestricted primary argument combinations. The viability of the method is shown by successfully extracting nested relationships in BioInfer and the corpus of the BioNLP'09 Shared Task on Event Extraction. The reported results, to the best of our knowledge, are the first for the nested relationships in BioInfer on a task in which only named entities are given.

1 Introduction

A recent shared task in biomedical text mining, the BioNLP'09 Shared Task on Event Extraction (Kim et al., 2009), showed that the biomedical natural language processing (BioNLP) community is greatly interested in heading towards the extraction of deep, semantically rich relationships. The shared task focused on biomolecular events involving proteins and called for methods that are capable of identifying nested structures. Biomolecular events are a major category of relationships in the biomedical domain in which, among others, relationships involving non-molecular entities such as diseases and static relations such as protein family memberships are also of interest.

Earlier, well-studied extraction tasks typically cast the problem in such a manner that relationships can be considered as mutually independent atomic units. However, as a nested semantic structure grows in its depth and in the total number of relationship arguments, its simultaneous extraction becomes difficult, if not impossible. Systems that bypass this problem by identifying atomic units of nested structures in a mutually independent manner must still decide which of the units collectively comprise a complete structure.

Another problem arises from the fact that a single syntactic token can refer to several, distinct relationships each having a unique combination of arguments. This is typically induced by coordinations which are common in the biomedical domain (Pyysalo et al., 2007). As a result, aside from the identification and classification of relationships and their potential arguments, extraction systems have to make decisions about how many relationships should be generated and how the arguments should be distributed among them. For example, the sentence "the binding of A and B to DNA regulates C and D, respectively" states that there are two binding events (*A–DNA* and *B–DNA*) the former of which regulates *C* and the latter *D* instead of, for example, that both binding events regulate both *C* and *D* or that there is a single binding event between *A*, *B*, and *DNA*.

This paper focuses on addressing the aforementioned problems in the case of the extraction method developed by Björne et al. (2010) for the BioNLP'09 Shared Task and generalises this method. Björne et al. showed that deep depen-

Proceedings of the 2010 Workshop on Biomedical Natural Language Processing, ACL 2010, pages 108–116,
Uppsala, Sweden, 15 July 2010.

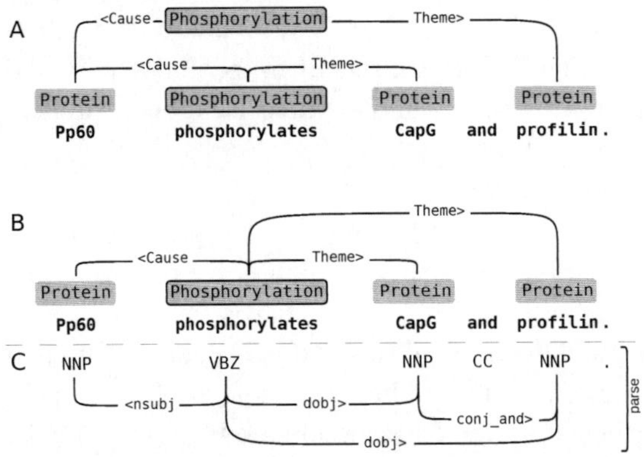

Figure 1: A one-node-per-token constrained graph (projected, B) cannot express the two distinct phosphorylation events while an unrestricted semantic graph (deprojected, A) can. A parse in the SD scheme is illustrated in C.

dency analyses in the well-established Stanford Dependency (SD) scheme (de Marneffe and Manning, 2008) can successfully be utilised in extracting graphs that express semantic entities as nodes and relationship arguments as edges but are limited to one node per syntactic token. Nodes and edges can be extracted in a mutually independent manner but the resulting graph cannot necessarily express all the real relationships. Rather, the graph can be seen as a projection of the original graph: each node and edge has been mapped to the representative of its equivalence class which is determined by the node and edge types and the referred tokens.

The research question of this paper is *can the original semantic graphs be reconstructed from projected graphs with an independent step in an information extraction (IE) process?* The objective of deprojection is illustrated as a transformation of the graph *B* to the graph *A* in Figure 1.

To answer the question, the problem of reconstructing complex, nested semantic structures from their projections is formulated and a generic deprojection method is proposed. The method specifically addresses primary arguments, as defined by the BioNLP'09 Shared Task, while leaving the extension to secondary arguments as a future work. The viability of the method is analysed with BioInfer (Pyysalo et al., 2007) and the BioNLP'09 Shared Task corpus, both of which containing nested structures, through an IE task essentially identical to the BioNLP'09 Shared Task. It is concluded that the proposed method

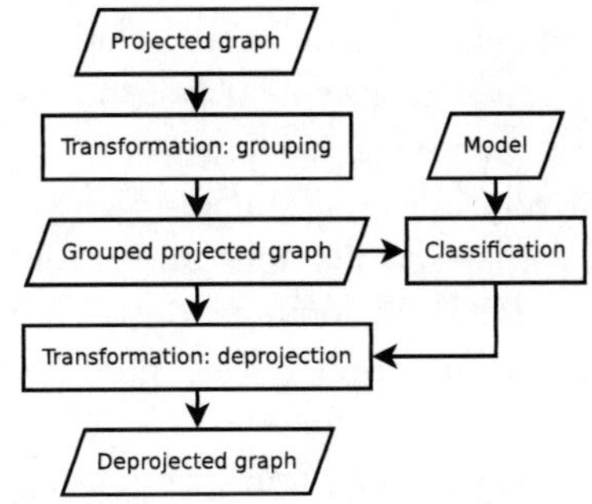

Figure 2: The deprojection process.

can successfully augment the method of Björne et al. (2010) and generalise it to arbitrary graphs of nested biomolecular relationships without the strict restrictions of the BioNLP'09 Shared Task while retaining its performance level. Thus, the method can improve IE systems that produce relationships on the one-per-token basis.

2 Method

The proposed approach to deproject semantic graphs is outlined in Figure 2. In summary, the first transformation (*grouping*) alters a projected graph such that a minimal set of classes is sufficient to describe the behaviour of the nodes and the edges. Guided by predicted class labels, the second transformation (*deprojection*) then produces a deprojected graph. In the presented method, the classification problem is solved with machine-learning (ML) methods. Finally, corpus-specific constraints are enforced.

2.1 Definitions

The graph representation of semantic annotation introduced by Heimonen et al. (2008) is adopted with some additional definitions. Semantic knowledge is represented as a directed acyclic graph (DAG) as follows.

Nodes and edges correspond to semantic entities (such as protein and processes) and relationship arguments, respectively. The equality of nodes is determined by the equality of their types and of their references to text. Similarly, the equality of edges arises from the equality of their types and of their end nodes.

Shallow and deep relationships consist of a node, its outgoing edges, and its direct successors. The latter also recursively include the successor relationships. Nodes are equal as shallow relationships if they as well as their outgoing edges are equal. Node equality as a deep relationship imposes the further requirement that the successors are equal as deep relationships.

A valid relationship is one which is valid in the given corpus-specific annotation scheme. Especially, it has a valid combination of arguments.

A deprojected graph (see Figure 1A) is one in which each node represents a valid, real relationship. Several equal nodes can exist provided they have unique combinations of outgoing edges. Note that there is one-to-one correspondence between nodes and real relationships but many-to-one between nodes and syntactic tokens.

A projected graph (see Figure 1B) is one generated by mapping each node and edge of a deprojected graph to the representative of its equivalence class. That is, each node represents a set of equal nodes of the deprojected graph, and similarly for edges. As a result, each token is referred to by at most one node[1] and there is a one-to-many correspondence between nodes and valid, real relationships. Also, the edges that are mapped to from the outgoing edges of equal nodes of the deprojected graph are the outgoing edges of a single node of the projected graph.

The deprojection of a semantic graph is the task of reproducing the original graph given a projected graph. This can also be seen as a task of finding the sets of outgoing edges that represent all the valid, real relationships.

2.2 Grouping

The objective of the first transformation is accomplished with a *grouping algorithm*: the direct successors of each node are grouped by their syntactic and semantic roles relative to the predecessor. The groups are represented as additional nodes in the graph. The rationale for this grouping is that similar arguments tend to either be mutually exclusive (and be associated with some other arguments) or together form a single relationship. This behaviour can easily be described with two classes: *distributive* and *collective*. For example, in the sentence "A and B regulate C", the entities *A* and *B* share both the argument type (*agent*) and the syntactic role (*subject*) relative to the relationship *regulate*. They form a group and are mutually exclusive (distributive) while this group forms a single relationship (collective) together with *C*. As a result, *A*–*C* and *B*–*C* pairs of regulation are generated. This approach relates to the collectivity and distributivity of plurals which have been studied, among others, by Scha and Stallard (1988) and Brisson (2003).

Technically, the grouping is a series of transformations in each of which a set of successors is replaced with a single, newly-created successor and the original successors become the successors of this node. The successors are first trivially grouped by the corresponding edge type. Finally, they are recursively grouped by syntactic similarity until they form a single group or multiple singleton groups. As a result, nested groups are generated.

The groups by syntax are determined by first mapping both the predecessor and the successors into the referred tokens in the syntactic graph. Then, the tokens referred to by the predecessor are removed if they are not also referred to by any of the successors. This removal step is recursively applied to the predecessors of the removed tokens. As a result, the syntactic graph is decomposed into several connected components, each of which representing a group. Thus, two successors are grouped if their referred tokens belong to the same connected component.

2.3 Deprojection

The second transformation is guided by node class labels (Figure 3). A collective node remains unchanged: its successors are kept together. In contrast, a distributive node is duplicated for each outgoing edge and the edges are distributed, one edge per duplicate. These node classes are enough to solve most of the cases in the analysed data sets. However, especially in BioInfer, this is not sufficient since the duplicates of a distributive node may themselves be either collective or distributive under their predecessor.

To adequately describe the behaviour of the duplicates generated by a single distributive node, the incoming edges of each distributive node are classified as *collective* or *distributive* (Figure 4). The duplication of a node also duplicates its incoming edges which are then processed by the assigned class labels as follows. In the case of a collective edge, the generated duplicates of the edge share the predecessor and are thus arguments in a single relationship. In contrast, a distributive edge induces the duplication of the predecessor re-

[1] given that, in the deprojected graph, a token can be referred to by multiple nodes only if they are of the same type

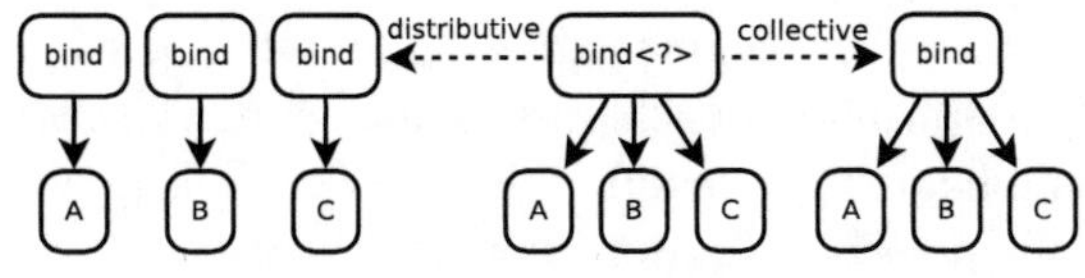

Figure 3: The effect of assigning collective or distributive class labels (marked as <?>) to a node in the deprojection process.

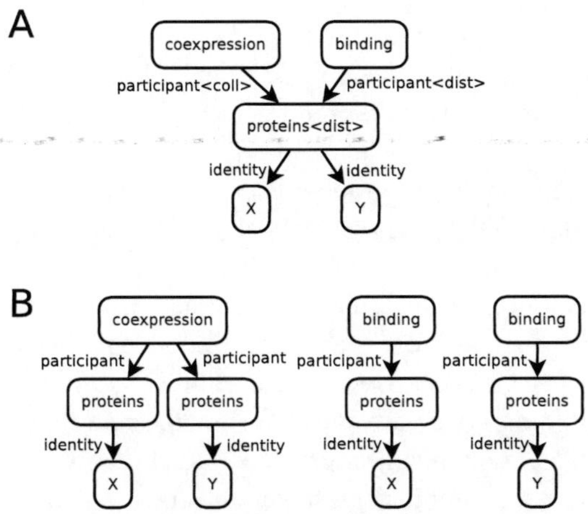

Figure 4: Correct node and edge class labels for the projected graph of the phrase "Coexpression and subsequent DNA-binding of X and Y proteins" (A) and the resulting deprojected graph (B).

lationships such that, as a result, the generated duplicate edges do not share any predecessors.

In Figure 4A, the node *proteins* is distributive because it represents two distinct nodes: one pertaining to *X*, another to *Y*. These two nodes are involved in the same *coexpression* relationship but in different *binding* relationships. Hence the incoming edges of the node *proteins* are collective and distributive, respectively.

Since the two transformation steps do not enforce corpus-specific constraints, a trivial algorithm is utilised to decompose relationships with invalid argument combinations into multiple valid relationships. In an ideal situation, this step makes no transformations. This is also used as a part of the baseline method (see Section 3.3).

2.4 Machine-learning and features

For node and edge classifications, the C4.5 decision tree (Quinlan, 1993), and its J48 implementation in the Weka package, was utilised because its models can easily be examined. This facilitates the analysis of the problem and the further development of the solution. The default parameters were used since no improvement was gained with alternative parameters in preliminary experiments. The applied feature set emphasises higher-level features obtained from the semantic and syntactic graphs. It consists of three main groups: semantic, syntactic, and morphological.

Semantic features contain information gathered from the semantic graph as well as from the type hierarchies. For nodes, these features consists of the node type as well as the presence, count and combination of outgoing edge types. The count of successor groups and the distance to the first non-group predecessor are also included. For edges, the node features are generated for both the successor and the predecessor in addition to the type of the edge.

Syntactic features include the minimum syntactic distance[2] from the predecessor to the successors as well as between the successors. Also, in the case of the unit distance, the corresponding dependency type is included.

Morphological features consist of the Porter stems (Porter, 1980) and the part-of-speech tags of the referred tokens as well as the presence and the Porter stems of the tokens that are shared between the successors.

All features are also generated from the first non-group predecessor (which may be the node itself) to capture the original relationship node when processing a group node. The majority of the features are Boolean-valued in order to allow several values of a single property. This is utilised in features representing hierarchical knowledge (such as node and dependency types) as well as stem features. For example, a node receives *true* for the node type feature of its actual type as well as of its supertypes in the hierarchy.

3 Resources and experiments

An array of experiments was performed to analyse the deprojection problem and the proposed solution. Firstly, the same experiments were performed on two corpora, BioInfer and the BioNLP'09 Shared Task corpus in order to evaluate the effect of the annotation scheme to the properties of the problem. Secondly, the deprojection algorithm was applied to both projected gold-standard graphs and to predicted graphs in order to study the effect of the accuracy of the input graph. Thirdly, the effect of the quality of the parse was

[2]semantic nodes mapped into the referred tokens

examined by employing various parses including the BioInfer gold-standard annotation.

3.1 Data

BioInfer is a corpus of 1100 sentences selected from 836 publication abstracts available through PubMed. For this paper, the abstracts were randomly sampled in the ratio 2:1:1 into the training, development, and test sets. In contrast, the BioNLP'09 Shared Task corpus consists of the training, development, and test sets of 800, 150, and 260 abstracts, respectively. Since the annotation of the test set is not publicly available and the evaluation server does not provide the required details for the analysis, the development set was used as the test set and a random sample of 150 abstracts was cut from the training set to form the development set.

In this study, the task 1 annotation with the protein equivalence relations removed was used as the BioNLP'09 Shared Task data set. In this annotation, relationships are positively asserted, have only *Theme* and *Cause* arguments and are annotated only for one of the equivalent proteins. Furthermore, each node refers to at least one token in the syntactic graph. The BioInfer semantic annotation was transformed into a similar form by removing negation (*NOT*), equivalence (*EQUAL*), and reference nodes (*COREFER*, *REL-ENT*). Furthermore, to create a fully text-bound subset, family memberships relations (*MEMBER*) were resolved into single edges and suitable references to text were added for the remaining unbound nodes when possible. In an extreme case, an unbound relationship was discarded. As a result, the differences to the BioNLP'09 Shared Task data set were minimised to additional node and edge types reflecting the wider selection of primary arguments.

All employed parses follow the SD scheme. BioInfer contains uncollapsed gold-standard parses while the BioNLP'09 Shared Task corpus includes parses, in the collapsed representation, generated by the parser of Charniak and Johnson (2005) using the model of McClosky and Charniak (2008). For both corpora, additional parses were produced with the improved version of the aforementioned system created by McClosky (2009). These parses were transformed into both the collapsed and the conjunct dependency propagated representations with the tools provided by de Marneffe et al. (2006). All parses were further augmented by splitting tokens at non-alphanumeric characters that border named entities and connecting the newly-created tokens with dependencies denoting the character.

3.1.1 Predicted graphs

The predicted semantic graphs were obtained as a result of an extraction task adopted from the BioNLP'09 Shared Task. In this task, named entities are given as gold-standard annotation and their relationships are to be extracted by identifying text spans, determining types, and assigning arguments.

The predicted graphs were produced with the system developed for the BioNLP'09 Shared Task by Björne et al. (2010). The system has two machine-learning steps. First, relationship nodes are predicted, one per token, based only on the syntax and the given named entity nodes. Next, outgoing edges are predicted for the relationship nodes. As a result, a projected graph is obtained. With the graph representation, the system can transparently be trained for both the BioNLP'09 Shared Task corpus and BioInfer regardless of the differences in their annotation schemes.

The two prediction steps utilise the SVM^{multiclass} implementation of a multiclass support vector machine (Crammer and Singer, 2002; Tsochantaridis et al., 2004). In this study, the steps were independently optimised for model parameters and, in contrast to the original training procedure, the recall boosting optimisation was omitted due to limited resources available. When training the edge prediction, the gold-standard relationship nodes were used.

In the graph prediction, the conjunct dependency propagated parses produced with the parser of McClosky (2009) were systematically applied.

3.2 Experiments

Original gold-standard graphs were used in generating decision tree models as well as subjected to projection. Predicted graphs and the projected gold-standard graphs were deprojected with the models. The evaluation of the deprojected graphs was performed against the original graphs.

During the system development, the training and development sets were available and the data were thoroughly analysed. The progress was estimated by training the system with the former and testing against the latter. The final results were obtained on the test sets by applying the system trained on the combined training–development set. For analysis, also the baseline method and the method of Björne et al. (2010) were evaluated on the test sets.

3.3 Baselines

The baselines were designed to reflect an IE system following the one-node-per-token principle without an advanced postprocessing but still enforcing the annotation scheme constraints.

With the strict specifications of the BioNLP'09 Shared Task, a sound baseline is obtained simply by enforcing the constraints through a minimal set of changes. Nodes with outgoing *Cause* and *Theme* edges are duplicated into all *Cause–Theme* pairs. *Binding* nodes remain unchanged since they can have several *Theme* arguments while the others are treated as distributive nodes with distributive incoming edges.

Although BioInfer is less restricted with respect to valid argument combinations, a feasible baseline can be obtained by adapting the BioNLP'09 Shared Task baseline algorithm. *Cause–Theme* is replaced with *agent–patient* while *Binding* is extended to symmetric relationships (i.e. *participant* arguments). In addition, relationships with *sub* arguments are treated as collective which reflects multiple components in a single complex. These changes were also applied to the method of Björne et al. (2010) when analysing BioInfer.

3.4 Evaluation

The standard precision–recall–F_1 metrics was used in the evaluation. True/false positive/negative instances were determined by the equality of the nodes as relationships: pairs of equal nodes were true positives while unique nodes in the deprojected and the original graph were false positives and false negatives, respectively.

The equality of references to text was determined after removing the tokens found in a non-exhaustive list of common stop-words including prepositions, articles, and non-alphanumeric characters. This relaxes an unnecessary requirement of the node prediction step to find also those tokens in the BioInfer annotation that do not contribute to the semantics of the nodes. For example, prepositions should be associated with edges rather than nodes.

The F_1-scores were further analysed with the Wilcoxon signed-rank test (Wilcoxon, 1945), as implemented in Scipy v. 0.7.0, by considering each document as an experiment and using the 95% confidence level.

BioInfer

method	gold total	gold symm.	predicted total	predicted symm.
baseline	88.26	63.62	29.38	18.64
Björne et al.	89.15	72.35	29.14	20.37
proposed	**92.42**	**78.79**	**30.79**	**24.47**

BioNLP'09

method	gold total	gold symm.	predicted total	predicted symm.
baseline	92.52	64.15	43.70	21.05
Björne et al.	94.51	83.37	45.13	35.21
proposed	**95.08**	**84.32**	**45.32**	**36.63**

Table 1: The F_1-scores on the test sets. *Total* is cumulative over all nodes with outgoing edges while *symm.* refers to the symmetric types. *Gold* and *predicted* refer to the experiments with gold-standard and predicted graphs, respectively.

4 Results and discussion

The following discussion focuses on the deep relationship equality as the evaluation criterion because it reflects the relationships of interest by requiring the identification of the pertaining named entities. Also, the discussion only considers the experiments performed with the conjunct dependency propagated parses obtained with the parser of McClosky (2009) because switching parses did not produce statistically significant differences in performance. Note that the results are not comparable to those of Björne et al. (2010) because the graph prediction was not fully optimised.

With respect to the deprojection task, BioInfer was found to be similar to the BioNLP'09 Shared Task corpus: it contains symmetric relationships (c.f. *Binding*), asymmetric relationships (c.f. *Regulation*), and single-argument relationships. Only the symmetric relationships are a challenge in the deprojection task because they can have an arbitrary number of arguments. In contrast, the baseline F_1-scores for the others are above 94% on the gold-standard graphs.

Table 1 shows the F_1-scores on the test sets of BioInfer and the BioNLP'09 Shared Task corpus for the overall performance as well as for the symmetric relationships only. The proposed method outperforms the two other methods in all experiments and the ΔF_1 against the proposed method are statistically significant with the exception of the method of Björne et al. (2010) on the BioNLP'09 Shared Task corpus. Although not conclusively better than the earlier, specialised method in its own task, the proposed method successfully achieves the intended generalisation without an adverse effect.

The observed improvement over the method of Björne et al. (2010) is likely due to two factors. First, using machine-learning rather than a simple rule-based system allows for more accurate modelling of the problem. Second, the proposed method can handle a wider variety of cases due to the classification of edges. For example, the graph in Figure 4A can correctly be deprojected, which is not possible for the earlier method. However, the latter factor is only effective on BioInfer, the more complex of the two corpora, which is consistent with the observed statistical significances.

The proposed deprojection method is currently limited to the phenomena encountered in the two analysed corpora since the decision to use binary classification was based on the experimental observation that neither class is appropriate only in rare cases. More classes will be needed to further generalise and improve the system. One such class could be *respective* which denotes a selective pairing of sibling nodes. For example, the sentence "A and B binds C and D, respectively" currently results in false positive pairs *A–D* and *B–C*. Similarly, adding secondary arguments (e.g. *location*) and relationship modifiers (e.g. *negation*) into consideration is likely to necessitate new, more complex transformations and their respective classes. Also, to filter out incorrectly predicted edges will require the introduction of additional classes. The critical question is whether a reasonably small set of classes with extensive enough a coverage can be found.

Another limitation is that the approach expects an annotation scheme in which relationship arguments have the tendency of following syntactic dependencies as observed for BioInfer by Björne et al. (2008). This expectation may deteriorate the performance on highly refined schemes which do not consider syntax. On the other hand, since it relies more on the syntactic than on the biological properties of the relationships, the proposed approach should be applicable beyond the domain of biomolecular events (e.g. to gene–disease relationships or static relations).

The F_1-scores in Table 1 indicate that the BioNLP'09 Shared Task corpus is easier to extract than BioInfer. This is likely due to the narrower scope and the stricter constraints of the former. In absolute terms, the proposed method yields the largest improvement over the baseline on the gold-standard graphs which suggests that it is negatively affected by the presence of false nodes/edges or that the predicted graphs contain relatively more relationships that are trivially deprojected. On the other hand, in relative terms, the largest improvements are observed for symmetric relationships in the BioNLP'09 Shared Task corpus but overall in BioInfer. This is likely due to the differences in the relationship type distributions.

The system recently developed by Miwa et al. (2010), based on the architecture of Björne et al. (2010), utilises a ML-based deprojection which enumerates all possible argument combinations and classifies them as positive or negative. While this approach may be prohibitively expensive in more complex schemes in which the number of arguments and their types is higher, it should outperform the proposed method on the BioNLP'09 Shared Task corpus. Since Miwa et al. do not analyse the contribution of the deprojection to the overall performance, a direct comparison of the two methods is impossible. In any case, the systems of Björne et al. (2010) and Miwa et al. (2010) demonstrate the success of the architecture using deprojection and further motivate the investigation of deprojection methods.

4.1 Future directions

In the future, the proposed method will be studied and further improved with two other corpora, GENIA Event Annotation (Kim et al., 2008) and Gene Regulation Event Corpus (Thompson et al., 2009), which are similar in their purpose compared to the already-analysed corpora. The former corpus is interesting because of the co-operativity of event participants which relaxes the restrictions on asymmetric relationships while the latter contains an extensive annotation for non-primary arguments. The method could also be examined with the static relation extraction task recently introduced by Pyysalo et al. (2009).

In addition to improving the method and extending it to non-primary arguments, embedding the presented approach to a joint inference system, such as Markov Logic Network (MLN), will be studied. Deprojection is likely to greatly benefit methods based on Markov Logic which is "not yet able to deal with cases where the number and identity of entities is unknown, while relations/links between known objects can be readily modelled" (Riedel et al., 2009). The objective is to combine the graph prediction and the deprojection steps as well as to simultaneously enforce task-specific constraints and adapt to the presence of false positive nodes and edges. This should be achievable by extending the methods developed for the BioNLP'09 Shared Task corpus by Riedel et al. (2009) or by Poon and Vanderwende (2010), both of which determine the correct argument combinations outside of the Markov Logic framework.

Semantic role labelling (SRL) is a task similar to the graph-based relationship extraction applied in this paper although the former typically only concerns shallow predicate–argument structures (Hacioglu, 2004; Surdeanu et al., 2008). The similarities between the tasks suggest that exploring them jointly may benefit the development of information extraction methods.

In the long term, semantic schemes should be developed such that, ideally, all syntactic tokens are considered for their semantics and semantic relationships readily follow from their dependencies. Such schemes, closely following the syntax, could improve both the graph prediction and the deprojection. In this research direction, graph-based knowledge representations such as conceptual graphs (Sowa, 1976; Chein and Mugnier, 2008) or graphical logical forms such as the one proposed by Allen et al. (2008) could be adopted.

Given the frequency of coordinations in the biomedical domain, deprojection may prove to be useful in the development of deep semantic parsing in the biomedical domain. For example, with improved semantic schemes, it could provide a means to generate complete, detailed semantic graphs directly from deep dependency analyses in a single-step by applying joint inference to achieve simultaneous node/edge relabelling and graph deprojection.

5 Conclusions

This study presents a method for reconstructing the original semantic graphs from their projections by determining the correct combinations of relationship arguments. It generalises the postprocessing step of the system described by Björne et al. (2010) and extends the extraction capability of this system to arbitrary graphs of nested biomolecular relationships. The evaluation of the method on BioInfer and the BioNLP'09 Shared Task corpus indicates that the approach is viable for primary relationship arguments. For BioInfer, the outcome is, to the best of our knowledge, the first reported result of the task of extracting the nested relationships in its original version.

The presented method facilitates an IE approach in which the identification of semantic entities is performed on the one-entity-per-token basis and relationship arguments are identified in a mutually independent manner disregarding the semantics of the argument combinations. The method handles the selection of the correct argument combinations, which is non-trivial particularly when coordinations are involved, and generates the final output in which a single token can refer to several entities. This approach improves the utilisation of deep dependency analyses by simplifying the correlation between them and semantic graphs. Due to its independent nature, the method can be coupled to any system identifying relationships on the one-per-token basis.

The implemented system will be available upon request.

Acknowledgements

Thanks to Filip Ginter for his help with the parses. This study was funded by Academy of Finland. Computational resources were provided by CSC – IT Center for Science.

References

J. Allen, M. Swift, and W. de Beaumont. 2008. Deep semantic analysis of text. In *Proceedings of the 2008 Conference on Semantics in Text Processing (STEP'08)*, pages 343–354.

J. Björne, S. Pyysalo, F. Ginter, and T. Salakoski. 2008. How complex are complex protein-protein interactions? In *Proceedings of the Third International Symposium on Semantic Mining in Biomedicine (SMBM'08)*, pages 125–128.

J. Björne, J. Heimonen, F. Ginter, A. Airola, T. Pahikkala, and T. Salakoski. 2010. Extracting contextualized complex biological events with rich graph-based feature sets. *Computational Intelligence*. To appear.

C. Brisson. 2003. Plurals, all, and the nonuniformity of collective predication. *Linguistics and Philosophy*, 26:129–184.

E. Charniak and M. Johnson. 2005. Coarse-to-fine n-best parsing and maxent discriminative reranking. In *Proceedings of the 43rd Annual Meeting of the Association for Computational Linguistics (ACL'05)*, pages 173–180.

M. Chein and M.-L. Mugnier. 2008. *Graph-based Knowledge Representation: Computational Foundations of Conceptual Graphs*. Springer Publishing Company Inc.

K. Crammer and Y. Singer. 2002. On the algorithmic implementation of multiclass kernel-based vector machines. *Journal of Machine Learning Research*, 2:265–292.

M.-C. de Marneffe and C. Manning. 2008. The stanford typed dependencies representation. In *Proceedings of the Coling'08 Workshop on Cross-Framework and Cross-Domain Parser Evaluation*, pages 1–8.

M.-C. de Marneffe, B. MacCartney, and C. Manning. 2006. Generating typed dependency parses from phrase structure parses. In *Proceedings of the Fifth International Conference on Language Resources and Evaluation (LREC'06)*, pages 449–454.

K. Hacioglu. 2004. Semantic role labeling using dependency trees. In *Proceedings of the 20th International Conference on Computational Linguistics (COLING'04)*, pages 1273–1276.

J. Heimonen, S. Pyysalo, F. Ginter, and T. Salakoski. 2008. Complex-to-pairwise mapping of biological relationships using a semantic network representation. In *Proceedings of the Third International Symposium on Semantic Mining in Biomedicine (SMBM'08)*, pages 45–52.

J.-D. Kim, T. Ohta, and J. Tsujii. 2008. Corpus annotation for mining biomedical events from literature. *BMC Bioinformatics*, 9:10.

J.-D. Kim, T. Ohta, S. Pyysalo, Y. Kano, and J. Tsujii. 2009. Overview of bionlp'09 shared task on event extraction. In *Proceedings of the NAACL–HLT'09 Workshop on BioNLP: Companion Volume for Shared Task (BioNLP'09)*, pages 1–9.

D. McClosky and E. Charniak. 2008. Self-training for biomedical parsing. In *Proceedings of the 46th Annual Meeting of the Association for Computational Linguistics: Human Language Technologies, Short Papers (ACL'08: HLT)*, pages 101–104.

D. McClosky. 2009. *Any Domain Parsing: Automatic Domain Adaptation for Natural Language Parsing*. Ph.D. thesis, Department of Computer Science, Brown University, Providence, Rhode Island, USA.

M. Miwa, R. Saetre, J.-D. Kim, and J. Tsujii. 2010. Event extraction with complex event classification using rich features. *Journal of Bioinformatics and Computational Biology*, 8:131–146.

H. Poon and L. Vanderwende. 2010. Joint inference for knowledge extraction from biomedical literature. In *Proceedings of Human Language Technologies: The 11th Annual Conference of the North American Chapter of the Association for Computational Linguistics (NAACL–HLT'10)*.

M. Porter. 1980. An algorithm for suffix stripping. *Program*, 14:130–137.

S. Pyysalo, F. Ginter, J. Heimonen, J. Björne, J. Boberg, J. Järvinen, and T. Salakoski. 2007. BioInfer: a corpus for information extraction in the biomedical domain. *BMC Bioinformatics*, 8:50.

S. Pyysalo, T. Ohta, J.-D. Kim, and J. Tsujii. 2009. Static relations: a piece in the biomedical information extraction puzzle. In *Proceedings of the NAACL–HLT'09 Workshop on BioNLP (BioNLP'09)*, pages 1–9.

J. Quinlan. 1993. *C4.5: programs for machine learning*. Morgan Kaufmann Publishers Inc.

S. Riedel, H.-W. Chun, T. Takagi, and J. Tsujii. 2009. A markov logic approach to bio-molecular event extraction. In *Proceedings of the NAACL–HLT'09 Workshop on BioNLP: Companion Volume for Shared Task (BioNLP'09)*, pages 41–49.

R. Scha and D. Stallard. 1988. Multi-level plurals and distributivity. In *Proceedings of the 26th Annual Meeting of the Association for Computational Linguistics (ACL'88)*, pages 17–24.

J. Sowa. 1976. Conceptual graphs for a data base interface. *IBM Journal of Research and Development*, 20:336–357.

M. Surdeanu, R. Johansson, A. Meyers, L. Màrquez, and J. Nivre. 2008. The CoNLL 2008 shared task on joint parsing of syntactic and semantic dependencies. In *Proceedings of the Twelfth Conference on Computational Natural Language Learning (CoNLL'08)*, pages 159–177.

P. Thompson, S. Iqbal, J. McNaught, and S. Ananiadou. 2009. Construction of an annotated corpus to support biomedical information extraction. *BMC Bioinformatics*, 10:349.

I. Tsochantaridis, T. Hofmann, T. Joachims, and Y. Altun. 2004. Support vector machine learning for interdependent and structured output spaces. In *Proceedings of the 21st International Machine Learning Conference (ICML'04)*, page 104.

F. Wilcoxon. 1945. Individual comparisons by ranking methods. *Biometrics Bulletin*, 1:80–83.

Towards Internet-Age Pharmacovigilance: Extracting Adverse Drug Reactions from User Posts to Health-Related Social Networks

Robert Leaman[1], Laura Wojtulewicz[2], Ryan Sullivan[2]
Annie Skariah[2], Jian Yang[1], Graciela Gonzalez[2]
[1]School of Computing, Informatics and Decision Systems Engineering
[2]Department of Biomedical Informatics
Arizona State University, Tempe, Arizona, USA
{robert.leaman, whitz, rpsulli, annie.skariah,
jian.yang, graciela.gonzalez}@asu.edu

Abstract

Adverse reactions to drugs are among the most common causes of death in industrialized nations. Expensive clinical trials are not sufficient to uncover all of the adverse reactions a drug may cause, necessitating systems for post-marketing surveillance, or pharmacovigilance. These systems have typically relied on voluntary reporting by health care professionals. However, self-reported patient data has become an increasingly important resource, with efforts such as MedWatch from the FDA allowing reports directly from the consumer. In this paper, we propose mining the relationships between drugs and adverse reactions as reported by the patients themselves in user comments to health-related websites. We evaluate our system on a manually annotated set of user comments, with promising performance. We also report encouraging correlations between the frequency of adverse drug reactions found by our system in unlabeled data and the frequency of documented adverse drug reactions. We conclude that user comments pose a significant natural language processing challenge, but do contain useful extractable information which merits further exploration.

1 Introduction

It is estimated that approximately 2 million patients in the United States are affected each year by severe adverse drug reactions, resulting in roughly 100,000 fatalities. This makes adverse drug reactions the fourth leading cause of death in the U.S, following cancer and heart diseases (Giacomini et al., 2007). It is estimated that $136 billion is spent annually on treating adverse drug reactions in the U.S., and other nations face similar difficulties (van Der Hooft et al., 2006; Leone et al., 2008). Unfortunately, the frequency of adverse drug reactions is often under-estimated due to a reliance on voluntary reporting (Bates et al., 2003; van Der Hooft et al., 2006).

While severe adverse reactions have received significant attention, less attention has been directed to the indirect costs of more common adverse reactions such as nausea and dizziness, which may still be severe enough to motivate the patient to stop taking the drug. The literature shows, however, that non-compliance is a major cause of the apparent failure of drug treatments, and the resulting economic costs are estimated to be quite significant (Urquhart, 1999; Hughes et al., 2001). Thus, detecting and characterizing adverse drug reactions of all levels of severity is critically important, particularly in an era where the demand for personalized health care is high.

1.1 Definitions

An adverse drug reaction is generally defined as an unintended, harmful reaction suspected to be caused by a drug taken under normal conditions (World Health Organization, 1966; Lee, 2006). This definition is sufficiently broad to include such conditions as allergic reactions, drug tolerance, addiction or aggravation of the original condition. A reaction is considered *severe* if it "results in death, requires hospital admission or prolongation..., results in persistent or significant disability/incapacity, or is life-threatening," or if it causes a congenital abnormality (Lee, 2006).

Proceedings of the 2010 Workshop on Biomedical Natural Language Processing, ACL 2010, pages 117–125,
Uppsala, Sweden, 15 July 2010.

1.2 Pharmacovigilance

The main sources of adverse drug reaction information are clinical trials and post-marketing surveillance instruments made available by the Food and Drug Administration (FDA), Centers for Disease Control and Prevention (CDC) in the United States, and similar governmental agencies worldwide. The purpose of a clinical trial, however, is only to determine whether a product is effective and to detect common serious adverse events. Clinical trials, by their nature and purpose, are focused on a limited number of participants selected by inclusion/exclusion criteria reflecting specific subject characteristics (demographic, medical condition and diagnosis, age). Thus, major uncertainties about the safety of the drug remain when the drug is made available to a wider population over longer periods of time, in patients with co-morbidities and in conjunction with other medications or when taken for off-label uses not previously evaluated.

Recently, the regulatory bodies of both the U.S. and the U.K. have begun programs for patient reporting of adverse drug reactions. Studies have shown that patient reporting is of similar quality to that of health professionals, and there is some evidence that patients are more likely to self-report adverse drug reactions when they believe the health professionals caring for them have not paid sufficient attention to an adverse reaction (Blenkinsopp et al., 2007). In general, however, the FDA advocates reporting only serious events through MedWatch.

Self-reported patient information captures a valuable perspective that might not be captured in a doctor's office, clinical trial, or even in the most sophisticated surveillance software. For this reason, the International Society of Drug Bulletins asserted in 2005 that "patient reporting systems should periodically sample the scattered drug experiences patients reported on the internet."

1.3 Social Networks

Social networks focusing on health related topics have seen rapid growth in recent years. Users in an online community often share a wide variety of personal medical experiences. These interactions can take many forms, including blogs, microblogs and question/answer discussion forums. For many reasons, patients often share health experiences with each other rather than in a clinical research study or with their physician (Davison et al., 2000). Such social networks bridge the geographical gap between people, allowing them to connect with patients who share similar conditions–something that might not be possible in the real world.

In this paper we propose and evaluate automatically extracting relationships between drugs and adverse reactions in user posts to health-related social network websites. We anticipate this technique will provide valuable additional confirmation of suspected associations between drugs and adverse reactions. Moreover, it is possible this technique may eventually provide the ability to detect novel associations earlier than with current methods.

2 Related Work

In the work closest in purpose to this study, two reviewers manually analyzed 1,374 emails to the BBC and 862 messages on a discussion forum regarding a link between the drug paroxetine and several adverse reactions including withdrawal symptoms and suicide (Medawara et al., 2002). The authors concluded that the user reports contained clear evidence of linkages that the voluntary reporting system then in place had not detected.

Not much work has been done to automatically extract adverse reactions from text, other than the SIDER side effect resource, which was created by mining drug insert literature (Kuhn et al., 2010). There is, however, significant literature support for mining more general concepts, such as diseases. MetaMap is a primarily lexical system for mapping concepts in biomedical text to concepts in the UMLS Metathesaurus (Aronson, 2001). The ConText system categorizes findings in clinical records as being negated, hypothetical, or historical (Harkema et al., 2009).

Most of the work on finding diseases concerns either biomedical text or clinical records. A notable exception is the BioCaster system, which detects infectious disease outbreaks by mining news reports posted to the web (Collier et al., 2008).

Health social networks have become a popular way for patients to share their health related experiences. A considerable amount of research has been devoted to this area (Moturu et al., 2008), but most of this work has focused on the study of social interactions and quality evaluation instead of text mining. Automated information extraction from health social network websites remains largely unexplored.

3 Data Preparation

We used the DailyStrength[1] health-related social network as the source of user comments in this study. DailyStrength allows users to create profiles, maintain friends and join various disease-related support groups. It serves as a resource for patients to connect with others who have similar conditions, many of whom are friends solely online. As of 2007, DailyStrength had an average of 14,000 daily visitors, each spending 82 minutes on the site and viewing approximately 145 pages (comScore Media Metrix Canada, 2007).

3.1 Data Acquisition

To efficiently gather user comments about specific drugs from the DailyStrength site, we implemented a highly parallelized automatic web crawler. All data was scraped from the raw HTML using regular expressions since the site has no open API. Users indicate a specific treatment when posting comments to DailyStrength, however we filter treatments which are not drugs. For each user comment we extracted the user ID, disease name, drug name, and comment text. While more information about each user is available at the site (gender, age, self-declared location, and length of membership at the site), we limited our data usage to just the comment data. The DailyStrength Privacy Policy states that comments made by users will be publicly available. All data was gathered in accordance with the DailyStrength Terms of Service, and to respect fair use the data will not be made publicly available without permission from the site.

3.2 Preparing the Lexicon

To enable finding adverse reactions in the user comments, we created a lexicon by combining terms and concepts from four resources.

The UMLS Metathesaurus is a resource containing many individual biomedical vocabularies (National Library of Medicine, 2008). We utilized a subset limited to the COSTART vocabulary created by the U.S. Food and Drug Administration for post-marketing surveillance of adverse drug reactions, which contains 3,787 concepts.

[1] http://www.dailystrength.org

The SIDER side effect resource contains 888 drugs linked with 1,450 adverse reaction terms extracted from pharmaceutical insert literature (Kuhn et al., 2010). We used the raw term found in the literature and the associated UMLS concept identifier (CUI).

The Canada Drug Adverse Reaction Database, or MedEffect[2], contains associations between 10,192 drugs and 3,279 adverse reactions, which we used to create a list of adverse reaction terms. We found many adverse reaction terms with very similar meanings, for example "appetite exaggerated," and "appetite increased," which we grouped together manually.

We also included a small set of colloquial phrases we located manually in a subset of the DailyStrength comments and mapped to UMLS CUIs. This list is available[3], and includes the terms "throw up," meaning *vomit*, "gain pounds," meaning *weight gain*, and "zonked out," meaning *somnolence*.

We considered all terms which are associated with the same UMLS concept identifier (CUI) as synonymous and grouped them into a single concept. We also merged all concepts containing a term in common into a single unified concept. Our lexicon contains 4,201 unified concepts, each containing between one and about 200 terms.

4 Annotation

We annotated comments relating to the following 4 drugs: carbamazepine, olanzapine, trazodone, and ziprasidone. These drugs were chosen because they are known to cause adverse reactions and we could verify our results with close collaborators. We retained but did not annotate comments for the drugs aspirin and ciprofloxacin; these comments are used during evaluation. Our data contains a total of 6,890 comment records. User comments were selected for annotation randomly and were independently annotated by two annotators.

Annotator 1 has a BS in biology, 10 years nursing experience in the behavioral unit of a long term care facility, and has dispensed all of the drugs annotated. Annotator 2 has a BS and an MS in neuroscience, and has work experience in data management for pharmaceutical-related clinical research and post-marketing drug surveillance.

[2] http://www.hc-sc.gc.ca/dhp-mps/medeff/index-eng.php

[3] http://diego.asu.edu/downloads/adrs

Concept	Definition
Adverse effect	A reaction to the drug experienced by the patient, which the user considered negative
Beneficial effect	A reaction to the drug experienced by the patient, which the user considered positive
Indication	The condition for which the patient is taking the drug
Other	A disease or reaction related term not characterizable as one of the above

Table 1: The concepts annotated in this study and their definitions.

4.1 Concepts Annotated

Each comment was annotated for mentions of *adverse effects*, *beneficial effects*, *indications* and *other* terms, as defined in table 1. Each annotation included the span of the mention and the name of the concept found, using entries from the lexicon described in section 3.2. Each annotation also indicates whether it refers to an *adverse effect*, a *beneficial effect*, an *indication* or an *other* term, which we shall call its characterization.

4.2 Annotation Practices

There are four aspects which require careful consideration when characterizing mentions. First, the stated concept may or may not be actually experienced by the patient; mentions of concepts not experienced by the patient were categorized as *other*. Second, the user may state that the concept is the reason for taking the drug. If so, the mention was categorized as an *indication*. Third, the concept may be an effect caused by the drug. In this case, the mention is categorized as either an *adverse effect* or a *beneficial effect* based on whether the user considers the effect a positive one. This requires some judgment regarding what people normally view as positive – while sleepiness is normally an *adverse effect*, someone suffering from insomnia would consider it a *beneficial effect*, regardless of whether insomnia is the primary reason for taking the drug. Mentions of concepts which were experienced by the patient but neither an effect of the drug nor the reason for taking it were also categorized as *other*. Concepts were characterized as an *adverse effect* unless the context indicated otherwise.

Comments not containing a mention or that only indicated the presence of an adverse effect ("Gave me weird side effects") were discarded. If more than one mention occurred in a comment, then each mention was annotated separately.

Some comments clearly mentioned an adverse reaction, but the reaction itself was ambiguous. For example, in the comment "It did the job when I was really low. However, I BALLOONED on it," the annotator could infer "BALLOONED" to mean either *weight gain* or *edema*. A frequent example is colloquial terms such as "zombie," which could be interpreted as a physiological effect (e.g. *fatigue*) or a cognitive effect (e.g. *mental dullness*). In such cases, each mention was annotated by using both the context of the mention and annotator's knowledge of the effects of the drug.

Spans were annotated by choosing the minimum span of characters from the comment that would maintain the meaning of the term. Locating the mention boundaries was straightforward in many cases, even when descriptive words were in the middle of the term ("It works better than the other meds ive taken but I am gaining some weight"). However some comments were not as simple ("it works but the pounds are packing on").

4.3 Corpus Description

A total of 3,600 comments were annotated, a sample of which can be seen in table 2. We reserved 450 comments for system development. The annotators found 1,260 *adverse effects*, 391 *indications*, 157 *beneficial effects* and 78 *other*, for a total of 1,886 annotations.

We measured the agreement between annotators by calculating both kappa (κ) (Cohen, 1960) and inter-annotator agreement (IAA). For κ, we considered agreement to mean that the concept terms were in the same unified concept from the lexicon and the characterization of the mentions matched, since there is no standard method for calculating κ which includes the span. For IAA, we added the constraint that the annotation spans must overlap, since discussions of IAA typically include the span. Using these definitions, κ was calculated to be 85.6% and IAA to be 85.3%[4].

5 Text Mining

Since the drug name is specified by the user when the comment is submitted to DailyStrength, no ex-

[4] κ>IAA here due to the different definitions of agreement.

Sample Comments	Annotations
hallucinations and weight gain	"hallucinations" - *hallucinations: adverse effect*; "weight gain" - *weight gain: adverse effect*
This has helped take the edge off of my constant sorrow. It has also perked up my appetite. I had lost a lot of weight and my doctor was concerned.	"constant sorrow" - *depression: indication*; "perked up my appetite" - *appetite increased: beneficial effect*; "lost a lot of weight" - *weight loss: other*
It worked well, but doctor didn't asked for the treatment to continue once my husband was doing well again.	*none*
ARGH! Got me nicely hypomanic for two weeks, then pooped out on me and just made me gain a half pound a day so I had to stop.	"hypomanic" - *hypomania: beneficial effect*; "pooped out" - *tolerance: adverse effect*; "gain a half a pound a day" - *weight gain: adverse effect*
Works to calm mania or depression but zonks me and scares me about the diabetes issues reported.	"mania" - *mania: indication*; "depression" - *depression: indication*; "zonks me" - *somnolence: adverse effect*; "diabetes" - *diabetes: other*
Works for my trigeminal neuralgia. Increasing to see if it helps stabalize mood. Fatigue!	"trigeminal neuralgia" - *trigeminal neuralgia: indication*; "stabalize mood" - *emotional instability: indication*; "Fatigue" - *fatigue: adverse effect*
Take for seizures and bipolar works well	"seizures" - *seizures: indication*; "bipolar" - *bipolar disorder: indication*
fatty patti!	"fatty" - *weight gain: adverse effect*

Table 2: An illustrative selection of uncorrected comments submitted to the DailyStrength health-related social networking website, and their associated annotations.

traction was necessary for drug names. To extract the adverse drug reactions from the user comments, we implemented a primarily lexical method, utilizing the lexicon discussed in section 3.2.

5.1 Methods Used

Each user comment was split into sentences using the Java sentence breaker, tokenized by splitting at whitespace and punctuation, and tagged for part-of-speech using the Hepple tagger (Hepple, 2000). Stop-words were removed from both user comments and lexical terms[5]. Tokens were stemmed using the Snowball implementation of the Porter2 stemmer[6].

Terms from the lexicon were found in the user comments by comparing a sliding window of tokens from the comment to each token in the lexical term. The size of the window is configurable and set to 5 for this study since that is the number of tokens in the longest term found by the annotators. Using a sliding window allows the tokens to be in different orders and for there to be irrelevant tokens between the relevant ones, as in *weight gain* and "gained a lot of weight."

Since user comments contain many spelling errors, we used the Jaro-Winkler measurement of string similarity to compare the individual tokens (Winkler, 1999). We scored the similarity between the window of tokens in the user comment and the tokens in the lexical term by pairing them as an assignment problem (Burkard et al., 2009). We then summed the similarities of the individual tokens and normalized the result by the number of tokens in the lexical term. This score is calculated for both the original tokens and the stemmed tokens in the window, and the final score is taken to be the higher of the two scores. The lexical term is considered to be present in a user comment if the final score is greater than a configurable threshold.

We noted that most mentions could be categorized by using the closest verb to the left of the mention, as in "taking for seizures." As this study focuses on *adverse effects*, we implemented a filtering method to remove *indications*, *beneficial effects*, and *other* mentions on a short list of verbs we found to indicate them. Verbs on this list include "helps," "works," and "prescribe" all of which generally denote *indications*. The complete list is available[7].

5.2 Text Mining Results

We first evaluated the system against the 3,150 annotated comments not reserved for system development. Because our purpose is to find adverse drug reactions, we limited our evaluation to *ad-*

[5]http://ir.dcs.gla.ac.uk/resources/linguistic_utils/stop_words
[6]http://snowball.tartarus.org
[7]http://diego.asu.edu/downloads/adrs

verse effects. We used a strict definition of true positive, requiring the system to label the mention with a term from the same unified concept as the annotators. The results of this study are 78.3% precision and 69.9% recall, for an f-measure of 73.9%.

Since the purpose of this study is to determine if mining user comments is a valid way to find adverse reactions, we ran our system on all available comments and compared the frequencies of adverse reactions found against their documented incidence. We calculated the frequency that each *adverse effect* was found in the user comments for each of the drugs studied in this experiment. We then determined the most commonly found adverse reactions for each drug and compared them against the most common documented adverse reactions for the drug. Since the four drugs we chose for annotation all act primarily on the central nervous system, we added aspirin and ciprofloxacin for this study. The results of this evaluation contain encouraging correlations that are summarized in table 3.

6 Discussion

6.1 Error Analysis

We performed an analysis to determine the primary sources of error for our extraction system. We randomly selected 100 comments and determined the reason for the 24 false positives (FPs) and 29 false negatives (FNs) found.

The largest source of error (17% of FPs and 55% of FNs) was the use of novel adverse reaction phrases ("liver problem") and descriptions ("burn like a lobster"). This problem is due in part to idiomatic expressions, which may be handled by creating and using a specialist lexicon. This problem might also be partially relieved by the appropriate use of semantic analysis. However, this source of error is also caused by the users deliberately employing a high degree of linguistic creativity ("TURNED ME INTO THE SPAWN OF SATAN!!!") which may require deep background knowledge to correctly recognize.

The next largest source of error was poor approximate string matching (46% of FPs and 17% of FNs). While users frequently misspelled words, making lexical analysis difficult, the approximate string matching technique used also introduced many FPs. We note that spelling unfamiliar medical terminology is particularly difficult for users. Correcting this important source of error will require improved modeling of the spelling errors made by users.

Ambiguous terms accounted for 8% of the FPs and 7% of the FNs. While this is frequently a problem with colloquial phrases ("brain fog" could refer to *mental dullness* or *somnolence*), there are some terms which are ambiguous on their own ("numb" may refer to *loss of sensation* or *emotional indifference*). These errors can be corrected by improving the analysis of the context surrounding each mention.

Surprisingly, miscategorizations only accounted for 4% of the FPs. This small percentage seems to indicate that the simple filtering technique employed is reasonably effective. However this source of error can be seen more prominently in the frequency analysis, as seen in table 3. For example, one of the most frequent effects found in comments about trazodone was *insomnia*, which is one of its most common off-label uses. Other examples included *depression* with olanzapine, *mania* with ziprasidone, and *stroke* with aspirin. We note that since conditions not being experienced by the patient are always categorized as *other*, our techniques should profit somewhat from an extension to handle negation.

6.2 Analysis of Documented vs. Found Adverse Reactions

The experiment comparing the documented incidence of adverse reactions to the frequency they are found contained some interesting correlations and differences. We begin by noting that the adverse reaction found most frequently for all 6 of the drugs corresponded to a documented adverse reaction. There were also similarities in the less common reactions, such as *diabetes* with olanzapine and *bleeding* with aspirin. In addition, many of the adverse reactions found corresponded to documented, but less common, reactions to the drug. Examples of this included *edema* with olanzapine, *nightmares* with trazodone, *weight gain* with ziprasidone, *tinnitus* with aspirin, and *yeast infection* with ciprofloxacin.

One interesting difference is the relative frequency of "hangover" in the comments for ziprasidone. Since the users were not likely referring to a literal *hangover*, they were probably referring to the *fatigue*, *headache*, *dry mouth* and *nausea* that accompany a *hangover*, all of which are doc-

Drug name (Brand name)	Primary Indications	Documented Adverse Effects (Frequency)	Adverse Effects Found in User Comments (Frequency)
carbamazepine (Tegretol)	epilepsy, trigeminal neuralgia	**dizziness**, **somnolence or fatigue**, unsteadiness, **nausea**, vomiting	**somnolence or fatigue** (12.3%), allergy (5.2%), weight gain (4.1%), rash (3.5%), depression (3.2%), **dizziness** (2.4%), tremor/spasm (1.7%), headache (1.7%), appetite increased (1.5%), **nausea** (1.5%)
olanzapine (Zyprexa)	schizophrenia, bipolar disorder	**weight gain** (65%), alteration in lipids (40%), **somnolence or fatigue** (26%), increased cholesterol (22%), **diabetes** (2%)	**weight gain** (30.0%), **somnolence or fatigue** (15.9%), appetite increased (4.9%), depression (3.1%), tremor (2.7%), **diabetes** (2.6%), mania (2.3%), anxiety (1.4%), hallucination (0.7%), edema (0.6%)
trazodone (Oleptro)	depression	**somnolence or fatigue** (46%), **headache** (33%), dry mouth (25%), **dizziness** (25%), nausea (21%)	**somnolence or fatigue** (48.2%), nightmares (4.6%), insomnia (2.7%), addiction (1.7%), **headache** (1.6%), depression (1.3%), hangover (1.2%), anxiety attack (1.2%), panic reaction (1.1%), **dizziness** (0.9%)
ziprasidone (Geodon)	schizophrenia	**somnolence or fatigue** (14%), **dyskinesia** (14%), nausea (10%), constipation (9%), **dizziness** (8%)	**somnolence or fatigue** (20.3%), **dyskinesia** (6.0%), mania (3.7%), anxiety attack (3.5%), weight gain (3.2%), depression (2.4%), allergic reaction (1.9%), **dizziness** (1.2%), panic reaction (1.2%)
aspirin	pain, fever, reduce blood clotting	nausea, vomiting, **ulcers**, **bleeding**, stomach pain or upset	**ulcers** (4.5%), sensitivity (3.8%), stroke (3.1%), bleeding time increased (2.8%), somnolence or fatigue (2.7%), malaise (2.1%), weakness (1.4%), numbness (1.4%), **bleeding** (1.0%), tinnitus (0.7%)
ciprofloxacin (Cipro)	bacterial infection	diarrhea (2.3%), vomiting (2.0%), **abdominal pain** (1.7%), headache (1.2%), restlessness (1.1%)	**abdominal pain** (8.8%), malaise (4.4%), nausea (3.8%), allergy (3.1%), somnolence or fatigue (2.5%), dizziness (1.9%), weakness (1.6%), tolerance (1.5%), rash (1.3%), yeast infection (1.1%)

Table 3: List of drugs included in the subset for analysis, with their indications and 5 most common adverse effects together with their frequency of incidence in adults taking the drug over the course of one year, as listed in the FDA online drug library, http://www.accessdata.fda.gov/scripts/cder/drugsatfda (some frequency data is not available). Also the 10 most frequent adverse effects found in the the DailyStrength data using our automated system. Correlations are highlighted in bold.

umented adverse reactions to the drug.

Users frequently commented on *weight gain* and *fatigue* while ignoring other reactions such as *increased cholesterol*. While this may be because users are more conscious of issues they can directly observe, this hypothesis would not explain why other directly observable reactions such as *nausea* and *constipation* are not always reported. Determining the general trends in the differences between clinical and user reports is an important area for future work.

6.3 Limitations

The present study has some limitations. We did not analyze the demographics of the users whose comments we mined, though it is likely that they are predominantly from North America and English-speaking. In future work we intend to expand the range of users and compare their demographics against clinical studies of adverse reactions. Also, the drugs we annotated operate primarily on the central nervous system and therefore have different adverse reaction profiles than would other drugs with substantially different mechanisms. While the inclusion of aspirin and ciprofloxacin does provide some evidence these techniques are more generally applicable, we also intend to expand the range of drugs studied in future work.

6.4 Opportunities for Further Study

In addition to our current classification for adverse reactions, there are additional dimensions along which each user comment could be studied. For example, many comments describe the degree of the adverse reaction, which can be straightforward ("extremely") or more creative ("like a pig"). Also, many users explicitly state whether they are still taking the drug, typically indicating whether their physician took them off or whether they took themselves off (non-compliance), and whether adverse reactions were the reason. User comments can also be categorized as medically non-descriptive ("I took one tablet and could'nt get out of bed for days and felt like I got hit by a truck"), somewhat medically descriptive ("My kidneys were not functioning properly"), or medically sound ("I ended up with severe leg swelling"). Comments also typically indicate whether the user is the patient or a caretaker by being phrased in either the first person or third person narrative. Finally, users also frequently describe whether they thought the benefits of the drug outweighed the adverse effects. We believe these additional dimensions represent a fertile area for further research.

7 Conclusion

In summary, we have shown that user comments to health related social networks do contain extractable information relevant to pharmacovigilance. We believe this approach should be evaluated for the ability to detect novel relationships between drugs and adverse reactions.

In addition to the improvements discussed in section 6, we plan in future work to increase the scale of the study (additional drugs, additional data sources, more user comments), improve the characterization of reactions using rule-based patterns, and evaluate the improved system with respect to all characterizations.

Acknowledgments

The authors would like to thank Dr. Diana Petitti for her early support and suggestions, Tasnia Tahsin for reviewing an earlier version, Skatje Myers for locating mergeable reaction concepts, and the anonymous reviewers for many useful suggestions. The authors are grateful for support from Science Foundation Arizona grant CAA 0277-08, the Arizona Alzheimers Disease Data Management Core under NIH Grant NIA P30 AG-19610, and the Arizona Alzheimers Consortium pilot grant.

References

Alan R. Aronson. 2001. Effective mapping of biomedical text to the UMLS Metathesaurus: the MetaMap program. In *Proceedings of the AMIA Symposium*, page 17. American Medical Informatics Association.

D.W. Bates, R.S. Evans, H. Murff, P.D. Stetson, L. Pizziferri, and G. Hripcsak. 2003. Detecting adverse events using information technology. *Journal of the American Medical Informatics Association*, 10(2):115–128.

A. Blenkinsopp, M. Wang, P. Wilkie, and P. A. Routledge. 2007. Patient reporting of suspected adverse drug reactions: a review of published literature and international experience. *British Journal of Clinical Pharmacology*, 63(2):148–156.

Rainer Burkard, Mauro Dell'Amico, and Silvano Martello. 2009. *Assignment Problems*. Society for Industrial and Applied Mathematics.

Jacob Cohen. 1960. A Coefficient of Agreement for Nominal Scales. *Educational and Psychological Measurement*, 20(1):37–46.

Nigel Collier, Son Doan, Ai Kawazoe, Reiko Matsuda Goodwin, Mike Conway, Yoshio Tateno, Quoc-Hung Ngo, Dinh Dien, Asanee Kawtrakul, Koichi Takeuchi, Mika Shigematsu, and Kiyosu Taniguchi. 2008. BioCaster: detecting public health rumors with a Web-based text mining system. *Bioinformatics*, 24(24):2940–2941.

comScore Media Metrix Canada. 2007. Key Measures Report - Health.

K. P. Davison, J. W. Pennebaker, and S. S. Dickerson. 2000. Who talks? The social psychology of illness support groups. *The American Psychologist*, 55(2):205–217.

K.M. Giacomini, R.M. Krauss, D.M. Roden, M. Eichelbaum, M.R. Hayden, and Y. Nakamura. 2007. When good drugs go bad. *Nature*, 446(7139):975–977.

Henk Harkema, John N. Dowling, Tyler Thornblade, and Wendy W. Chapman. 2009. ConText: An algorithm for determining negation, experiencer, and temporal status from clinical reports. *Journal of Biomedical Informatics*, 42(5):839851.

Mark Hepple. 2000. Independence and commitment: Assumptions for rapid training and execution of rule-based POS taggers. In *Proceedings of the 38th Annual Meeting of the Association for Computational Linguistics*, pages 277–278.

Dyfrig A. Hughes, Adrian Bagust, Alan Haycox, and Tom Walley. 2001. The impact of non-compliance on the cost-effectiveness of pharmaceuticals: a review of the literature. *Health Economics*, 10(7):601–615.

International Society Of Drug Bulletins. 2005. Berlin Declaration on Pharmacovigilance.

Michael Kuhn, Monica Campillos, Ivica Letunic, Lars Juhl Jensen, and Peer Bork. 2010. A side effect resource to capture phenotypic effects of drugs. *Molecular Systems Biology*, 6:343–348.

Anne Lee, editor. 2006. *Adverse Drug Reactions*. Pharmaceutical Press, second edition.

Roberto Leone, Laura Sottosanti, Maria Luisa Iorio, Carmela Santuccio, Anita Conforti, Vilma Sabatini, Ugo Moretti, and Mauro Venegoni. 2008. Drug-Related Deaths: An Analysis of the Italian Spontaneous Reporting Database. *Drug Safety*, 31(8):703–713.

Charles Medawara, Andrew Herxheimer, Andrew Bell, and Shelley Jofre. 2002. Paroxetine, Panorama and user reporting of ADRs: Consumer intelligence matters in clinical practice and post-marketing drug surveillance. *The International Journal of Risk and Safety in Medicine*, 15(3):161169.

S. T. Moturu, H. Liu, and W. G. Johnson. 2008. Trust evaluation in health information on the World Wide Web. In *30th Annual International Conference of the IEEE Engineering in Medicine and Biology Society*, pages 1525–1528.

National Library of Medicine. 2008. UMLS Knowledge Sources.

John Urquhart. 1999. Pharmacoeconomic consequences of variable patient compliance with prescribed drug regimens. *PharmacoEconomics*, 15(3):217–228.

Cornelis S. van Der Hooft, Miriam C. J. M. Sturkenboom, Kees van Grootheest, Herre J. Kingma, and Bruno H. Ch. Stricker. 2006. Adverse drug reaction-related hospitalisations: a nationwide study in The Netherlands. *Drug Safety*, 29(2):161–168.

William E. Winkler. 1999. The state of record linkage and current research problems.

World Health Organization. 1966. International Drug Monitoring: The Role of the Hospital.

Semantic role labeling of gene regulation events: preliminary results

Roser Morante
CLiPS - University of Antwerp
Prinsstraat 13, B-2000 Antwerpen, Belgium
Roser.Morante@ua.ac.be

Abstract

This abstract describes work in progress on semantic role labeling of gene regulation events. We present preliminary results of a supervised semantic role labeler that has been trained and tested on the GREC corpus.

1 Introduction

Semantic role labeling (SRL) is a natural language processing task that consists of identifying the arguments of predicates within a sentence and assigning a semantic role to them. This task can support the extraction of relations from biomedical texts. Recent research has produced a rich variety of SRL systems to process general domain corpora. However, only a few systems have been developed to process biomedical corpora (Tzong-Han Tsai et al, 2007; Bethard et al., 2008). In this abstract, we present preliminary results of a new system that is trained on the GREC corpus (Thompson et al., 2009).

The GREC corpus consists of 240 MEDLINE abstracts, in which gene regulation events have been annotated with different types of information, like the span of the event and of its arguments, and the semantic role of the arguments. Events can be verbs (58%) and nominalised verbs (42%). The corpus is divided into two species-specific subcorpora: *E. coli* (167 abstracts, 2394 events) and human (73 abstracts, 673 events).

2 System description

We perform two preprocessing steps. First, we extract the text and parse it with the GDep parser (Sagae and Tsujii, 2007) and then we convert the corpus from xml into CoNLL format. Table 1 shows a preprocessed sentence. The system performs argument identification and semantic role assignment in a single step, assuming gold standard event identification. It consists of one classifier that classifies an instance into one of the semantic role classes or the NONE class. An instance represents a combination of an event and a potential argument (PA). In order to generate the PAs, the system relies on information from the dependency syntax tree, which means that errors in the syntactic tree influence directly the performance of the system. We consider that the following tokens or combinations of tokens can be PAs: main verbs, nouns, adjectives, pronouns and adverbs; main verbs, nouns, adjectives, pronouns and adverbs with their modifiers to the left in the string of words; main verbs, nouns, adjectives, pronouns, adverbs, prepositions and relative pronouns with their modifiers to the left and to the right in the string of words.

The features extracted to perform the classification task are the following:

• About the event and the PA: chain of words, lemmas, POS, and dependency labels of all the tokens; lemma, POS and dependency label of head token, first token and last token; lemma and POS of syntactic father of head; lemma, POS, and dependency label of previous and next three tokens in the string of words; even type.

• About the dependency tree: feature indicating who is the ancestor (event, PA, other); lemma, POS, and dependency label of the first common ancestor of event and PA, if there is one; chain of dependency labels and chain of POS from event to common ancestor, and from PA to common ancestor, if there is one; chain of dependency labels and chain of POS from PA to event, if event is ancestor of PA; chain of dependency labels and chain of POS from event to PA, if PA is ancestor of event; chain of dependency labels and POS from event to ROOT and from PA to ROOT.

• Normalised distance in number of tokens between event an potential argument in the string of words.

We use an IB1 memory–based algorithm as implemented in TiMBL (version 6.1.2) [1](Daelemans et al., 2009), a memory-based classifier based on the k-nearest neighbor rule. The IB1 algorithm was parameterised by using Jeffrey divergence as the similarity metric, gain ratio for feature weighting, using 5 k-nearest neighbors, and weighting

[1]TiMBL: http://ilk.uvt.nl/timbl

Proceedings of the 2010 Workshop on Biomedical Natural Language Processing, ACL 2010, pages 126–127,
Uppsala, Sweden, 15 July 2010. ©2010 Association for Computational Linguistics

#	WORD	LEMMA	CHUNK	POS	DEP	LABEL	#E	TYPE		ROLES	
1	Lrp	Lrp	B-NP	NN	2	SUB	-	-	B-Agent	B-Agent	B-Agent
2	binds	bind	B-VP	VBZ	0	ROOT	E1	GRE	-	-	-
3	to	to	B-PP	TO	2	VMOD	-	-	-	-	-
4	two	two	B-NP	CD	5	NMOD	-	-	-	-	-
5	regions	region	I-NP	NNS	3	PMOD	-	-	-	-	-
6	in	in	B-PP	IN	5	NMOD	-	-	-	-	-
7	the	the	B-NP	DT	10	NMOD	-	-	B-Destination	-	-
8	dadAX	dadAX	I-NP	NN	10	NMOD	-	-	I-Destination	-	-
9	promoter	promoter	I-NP	NN	10	NMOD	-	-	I-Destination	-	-
10	region	region	I-NP	NN	6	PMOD	-	-	I-Destination	-	-
11	of	of	B-PP	IN	10	NMOD	-	-	-	-	-
12	Escherichia	Escherichia	B-NP	FW	13	NMOD	-	-	-	-	-
13	coli	coli	I-NP	FW	11	PMOD	-	-	-	-	-
14	to	to	B-VP	TO	15	VMOD	-	-	-	-	-
15	repress	repress	I-VP	VB	13	NMOD	E2	Gene_Repression	-	-	-
16	and	and	I-VP	CC	15	VMOD	-	-	-	-	-
17	activate	activate	I-VP	VB	15	VMOD	E3	Gene_Activation	-	-	-
18	transcription	transcription	B-NP	NN	17	OBJ	-	-	-	B-Theme	B-Theme
19	directly	directly	B-ADVP	RB	17	VMOD	-	-	-	B-Manner	B-Manner
20	.	.	O	.	2	P	-	-	-	-	-

Table 1: Sentence 1 from abstract 10216857 in E. coli corpus. Column # contains the token number; WORD, the word; LEMMA to LABEL contain information provided by the GDEP parser; #E, the event number; TYPE, the type of event, and ROLES contains columns with argument labels for each event following textual order, i.e., the first column corresponds to the first event in #E, the second column to the second event, etc.

the class vote of neighbors as a function of their inverse distance.

3 Preliminary results

We provide 5 fold cross-validation (CV) and cross-domain (CD) results in Table 2. The CV results are obtained by training and testing on different partitions of the same corpus. The CD results are obtained by training on one corpus and testing on the other. Although we cannot directly compare this results with results of other systems on exactly the same corpus, Sasaki et al. (2008) report CV results on a corpus of 677 MEDLINE abstracts on E. Coli gene regulation events. The precision achieved by their system is 49.00 and the recall 18.60. We consider that the results of our system are encouraging to proceed with further research.

Corpus	Precision	Recall	F1
E coli CV	59.72	32.29	41.92
E coli CD	49.87	18.07	26.53
Human CV	47.98	22.43	30.57
Human CD	56.57	25.90	35.53

Table 2: F1, precision and recall for argument identification and labeling.

4 Future work

Future work will deal with incorporating domain specific knowledge and with improving the machine learning techniques. We will experiment with other algorithms, like Conditional Random Fields, which are well known sequence labelers. Additionally, we will implement also a constraint satisfaction algorithm.

Acknowledgments

This preliminary study was made possible through financial support from the University of Antwerp (GOA project BIOGRAPH).

References

S. Bethard, Z. Lu, J.H. Martin, and L. Hunter. 2008. Semantic role labeling for protein transport predicates. *BMC Bioinformatics*, 9:277.

W. Daelemans, J. Zavrel, K. Van der Sloot, and A. Van den Bosch. 2009. TiMBL: Tilburg memory based learner, version 6.2, reference guide. Technical Report Series 09-01, ILK, Tilburg, The Netherlands.

K. Sagae and J. Tsujii. 2007. Dependency parsing and domain adaptation with LR models and parser ensembles. In *Proc. of CoNLL 2007: Shared Task*, pages 82–94, Prague, Czech Republic.

Y. Sasaki, P. Thompson, Ph. Cotter, J. McNaught, and S. Ananiadou. 2008. Event frame extraction based on a gene regulation corpus. In *Proc. of Coling 2008*, pages 761–768, Manchester, UK.

P. Thompson, S. A Igbal, J. McNaught, and S. Ananiadou. 2009. Construction of an annotated corpus to support biomedical information extraction. *BMC Bioinformatics*, 10:349.

R. Tzong-Han Tsai et al. 2007. BIOSMILE: A semantic role labeling system for biomedical verbs using a maximum-entropy model with automatically generated template features. *BMC Bioinformatics*, 8:325.

Ontology-Based Extraction and Summarization of Protein Mutation Impact Information

Nona Naderi and René Witte
Department of Computer Science and Software Engineering
Concordia University, Montréal, Canada

1 Introduction

NLP methods for extracting mutation information from the bibliome have become an important new research area within bio-NLP, as manually curated databases, like the Protein Mutant Database (PMD) (Kawabata et al., 1999), cannot keep up with the rapid pace of mutation research. However, while significant progress has been made with respect to mutation detection, the automated extraction of the *impacts* of these mutations has so far not been targeted. In this paper, we describe the first work to automatically summarize impact information from protein mutations. Our approach is based on populating an OWL-DL ontology with impact information, which can then be queried to provide structured information, including a summary.

2 Background

Mutations are alterations, rearrangements, or duplications of genetic material, impacting protein properties like stability or activity. For example:

H86A/E/F/K/Q/W decreased the enzyme stability at 60° C by up to 95% and the transition temperature by 2.5° C to 5.8° C.

Impacts are described through other concepts, since mutational events may cause changes to physical quantities such as *pH* and *temperature*. As presented in the above example, the named mutations (H86A/E/F/K/Q/W) made changes to the thermostability by 2.5–2.8 degrees Celsius. Hence, we extract (i) units of measurement, e.g., *%*, *degree Celsius*, *min*; (ii) protein properties: *stability*, *activity* and others; and (iii) impact words, including *increase*, *stabilize*, and *reduce*.

Measurable impacts can thus be classified based on the type of effect (increase, decrease or destabilize) on the protein property.

3 Related work

Little previous work exists on automatically detecting and extracting mutation impacts. An exception is EnzyMiner (Yeniterzi and Sezerman, 2009), which performs document classification for disease-related mutations. This work differs significantly from ours, as we are concerned with sentence-level impact detection and summarization.

4 Mutation Impact Detection

Our main contribution for impact detection and summarization consists of two major parts: an ontology describing impacts on a semantic level, and an NLP pipeline for detecting impacts in documents in order to populate the ontology. Further analysis, including summarization, can then be performed on this NLP-populated ontology through ontology queries and reasoning.

Ontology Design. Our *Mutation Impact Ontology* conceptualizes impacts and the mutations associated with them. The main concepts are: Mutation: An alteration or a change to a gene and developing a different offspring. UnitOfMeasurement: A class for measurement units. MutImpact: Mutation effect on protein properties. ProteinProperty: A class for properties of "Protein" and subclassed by different properties like "Activity" and "Stability." To design the Mutation Impact Ontology, information about several other elements is needed: Text elements, biological entities and entity relations. The relations between these entities are expressed as OWL object properties.

Mutation Impact Extraction. Impacts are detected through a combination of an *OntoGazetteer* annotating impact words, measurement units, etc., and JAPE grammar transducers, e.g.:

```
Rule: MutationImpact
({Lookup.majorType == "onto_impact"}):impact --> {
try {
  // get Impact annotations
  gate.AnnotationSet impactSet = (gate.AnnotationSet)bindings.get("impact");
  ...
}
```

Here, the impact word that is marked as "Lookup" with a feature of "majorType," "onto_impact" is

Proceedings of the 2010 Workshop on Biomedical Natural Language Processing, ACL 2010, pages 128–129,
Uppsala, Sweden, 15 July 2010.

annotated as "MutImpact." Accordingly, "ProteinProperty" and "UnitOfMeasurement" are annotated through similar JAPE grammars. Finally, each sentence is annotated as containing impact information or not. All the units of measurement and protein properties (ProteinProperty) existing in that sentence (impact) are recorded for subsequent ontology export.

Mutation-Impact Relation Extraction. When the entities such as *mutations* and *impacts* are identified and annotated, the sentence containing the impact word expressions (MutImpact) is associated with the nearest "Mutation," making the simple assumption that the nearest mutation invokes the impacts mentioned. The complete sentence is then considered as an impact sentence.

For each mutation-impact relation, we record the connection together with a number of properties, including units of measurement and effects.

Ontology Population. After preprocessing the documents and extracting the entities, the ontology is populated with the extracted entities such as *mutations*, *mutation impact* and their relations *mutation impact relations*.

5 Impact Summarization

The exported, populated OWL impact ontology can be queried using the SPARQL query language. To summarize impacts for a certain mutation, we can simply query the ontology for all detected impacts and extract the corresponding impact sentences:

```
PREFIX onto: <http://www.owl−ontologies.com/unnamed.owl#>
SELECT ?sentence
FROM <http://www.owl−ontologies.com/unnamed.owl#>
WHERE { ?document onto:containsSentence ?sentence.
            ?sentence onto:contains ?MutImpact.
            ?Mutation  onto:mutationMutImpactRel ?MutImpact}
ORDER BY DESC (?document) DESC (?Mutation)
```

These are then collected into a textual summary providing the mutations with their impacts for the user, as shown in Fig. 1.

PMID 10860737	
Mutation	Impacts
N35D	As predicted from sequence comparisons, substitution of this asparagine residue with an aspartic acid residue (N35D BCX) shifts its pH optimum from 5.7 to 4.6, with an 20 % increase in activity. . .

PMID 8855954	
Mutation	Impacts
E123A	Mutation of a third conserved active site carboxylic acid (E123A) resulted in rate reductions of up to 1500-fold on poorer substrates,...
E127A	Elimination of the acid/base catalyst (E127A) yields a mutant for which the deglycosylation step is slowed some 200-300-fold as a consequence of removal of general base catalysis, but with little effect on the transition state structure at the anomeric center. Effects on the glycosylation step due to removal of the acid catalyst depend on the aglycon leaving group ability, with minimal effects on substrates requiring no general acid catalysis but large (>105-fold) effects on substrates with poor leaving groups...
. . .	. . .

Figure 1: Impact Summaries (Excerpts)

6 Evaluation

The performance of the system was evaluated on the abstracts of four different mutation corpora, each on a specific protein family: *Xylanase* (19 documents), *Haloalkane Dehalogenase* (23 documents), *Subtilisin* (5 documents), and *Dioxygenase* (11 documents). Altogether, 58 documents were manually annotated with their impacts. For each annotation "Sentence," a binary feature "impact" is considered. As long as an impact exists in the sentence, the feature "impact" is set to "Yes;" otherwise to "No." The results are shown in the Table below; here, #C, #P, #M, and #S correspond to the correct, partially correct, missing, and spurious impact sentences, respectively; and P, R, F are the precision, recall, and F-measure:

Impact detection evaluation results on four corpora							
Corpus	**#C**	**#P**	**#M**	**#S**	***P***	***R***	***F***
Haloalkane D.	171	2	24	22	0.882	0.873	0.877
Xylanase	140	2	19	17	0.886	0.875	0.881
Dioxygenase	77	0	13	14	0.846	0.855	0.850
Subtilisin	32	2	9	10	0.750	0.767	0.758

The evaluation of associating the mutations with their impacts has so far been performed on the "Xylanase" corpus:

	Precision	Recall	F-Measure
Lenient (Partial matches included)	88%	80%	91%
Average (of Lenient and Strict)	86%	76%	80%
Strict (Partial matches not counted)	51.8%	46.6%	49.06%

7 Discussion

Our Mutation Impact Ontology models mutation impacts in the biomedical domain, linking them to the texts where they are found. Although the detection of mutation impacts has shown to be successful by this simple proximity heuristic to some extent, in some cases the impacts are missing or detected partially. Also, in cases where the impacts caused by a set of mutations, just one mutation (the nearest one) is considered, and the remaining mutations are ignored. Impacts are not always the result of the nearest mutation; However, automatically analysing the text and specifying the correct mutation associated with the impacts needs more complex analysis.

References

T. Kawabata, M. Ota, and K. Nishikawa. 1999. The Protein Mutant Database. *Nucleic Acids Res*, 27(1):355–357.

S. Yeniterzi and U. Sezerman. 2009. Enzyminer: automatic identification of protein level mutations and their impact on target enzymes from pubmed abstracts. *BMC Bioinformatics*, 10(Suppl 8):S2.

Extracting Distinctive Features of Swine (H1N1) Flu through Data Mining Clinical Documents

Heekyong Park
Department of Biomedical Engineering,
Seoul National University
Seoul, South Korea.
care01@snu.ac.kr

Jinwook Choi, MD, PhD
Department of Biomedical Engineering,
Seoul National University
Seoul, South Korea.
jinchoi@snu.ac.kr

Abstract

Early recognition of distinguishing patterns of a novel pandemic disease is important. We introduce a methodological approach based on popular data mining techniques to extract key features and temporal patterns of swine (h1n1) flu that is discriminated from swine flu like symptoms.

1 Introduction

Early recognition of a novel pandemic is desirable to minimize its dissemination. However, it usually spends some time in developing a diagnostic test and people tend to omit the test for various reasons including cost, which leads to late recognition. Under these circumstances, symptoms and signs in clinical documents might be valuable indicators to have a population perspective on pandemic severity.

In this paper, we propose a methodological approach to extract features of swine(h1n1) flu distinctive to swine flu like patients' one.

2 Method

2.1 Data

We randomly selected twenty clinical documents from first visit records of patients who had visited emergency room in Seoul National University Hospital (SNUH) with suspected case of swine flu. Ten of the documents are RT-PCR test positive cases, which mean that the patient is swine flu infected patient, while the ten remaining documents are RT-PCR negative cases. Each document contains a patient's symptoms, observations, recent clinical histories, clinical plans, and diagnoses in natural language. The symptoms are mostly about upper respiratory infection related ones but some of the sample documents describe about ones related to other diseases.

2.2 Hypotheses

We hypothesized two things as follows. 1) We will be able to extract distinctive symptom set between swine flu and swine-flu like patients by adopting apriori association rule mining method. 2) Although the two target groups accompany similar symptoms, we will be able to make selected symptoms more discriminative by developing impact score and considering temporal aspects, development rate.

2.3 Distinguishing Symptoms Extraction

We modeled each clinical document as one *(tag, item$_1$, ..., item$_n$)* transaction of which *tag* and *items* indicate RT-PCR result and candidate features, respectively. We set symptoms, signs, travel, and contact information as candidate features. We divided target data into three groups (Table 1) and ran apriori association rule algorithm to produce rules associated with RT-PCR(+) or RT-PCR(-) cases. Then the items appeared in association rules are collected. The items are grouped into four sets according to their original rules and training data (Table 2). The union of I_{pos} and I_{pos_co} was regarded as a distinguishing feature set, I_f. To enhance discrimination ability, we used some weights to score them as shown in Table 2 and 3.

Input data set	Description
Data 1	10 h1n1 positive transactions
Data 2	10 h1n1 negative transactions
Data 3	10 h1n1 positive transactions + 10 h1n1 negative transactions

Table 1. Input data

Proceedings of the 2010 Workshop on Biomedical Natural Language Processing, ACL 2010, pages 130–131,
Uppsala, Sweden, 15 July 2010. ©2010 Association for Computational Linguistics

Input data	Association rule descendant	Unique item set in association rules	Weight
Data 1	H1n1 (+)	I_{pos}	4
Data 3	H1n1 (+)	I_{pos_co}	5
Data 2	H1n1 (-)	I_{neg}	-1
Data 3	H1n1 (-)	I_{neg_co}	-2

* Subscript pos means the items are selected from h1n1(+) association rules (e.g., pos <- fever cough dyspnea) and neg means opposite cases (e.g., neg <- myalgia chilling fever). co indicates the items are selected from rules with h1n1(+) and h1n1(-) cases mixed training data (Data 3).

Table 2. Feature sets and weights

Features	I_{pos}	I_{pos_co}	I_{neg}	I_{neg_co}	score
Dyspnea	O	O			9
Pharyngeal injection	O	O			9
Sore throat	O	O		O	7
Travel	O	O		O	7
Cough	O	O	O	O	6
Fever	O	O	O	O	6
Sputum	O	O	O	O	6
Cvat (costovertebral angle tenderness)		O			5
Rhinorrhea		O			5

Table 3. Distinctive feature set(I_f) of h1n1 cases accompanied with impact scores

2.4 Disease Development Pattern Analysis

We extracted temporal information of selected features in sample documents and modeled as interval constraints. We ran Floyd-Warshall's all-pairs-shortest path algorithm to get hidden temporal relationships between two features. We traced start time gaps of the features to compare temporal patterns of the development rate of external indicators between two data sets, Data1 and Data2.

3 Result

For input data, we produced transactions and initial temporal constraint network manually. We extracted distinctive feature set, applied our scoring method, and compared temporal aspects.

We limited support value threshold as 30% for Ipos and Ineg and 0% for Ipos_co and Ineg_co due to small data size. Eight signs and symptoms as well as travel information were extracted from 47 items as distinguishing features of swine flu (Table 3). Besides sputum, pharynx injection, cvat, and travel information, five symptoms are the ones contained in the latest Centers for Disease Control (CDC) H1N1 influenza case report form. The others are strong indicators as well. The previous version of the CDC form contains sputum in the signs and symptoms section, and Himmerick (2009) described that pharyngeal injection is a clinical sign of uncomplicated swine-origin influenza A. Travel information is another important factor to diagnose an h1n1 case in Korea and included in a special purpose h1n1 clinical document format in SNUH emergency department.

We compared start time of the features but could not find any differences. The symptoms had been developed so rapidly that temporal pattern comparison between two groups was not meaningful. All the selected symptoms were developed within three days, and moreover, the symptoms in twelve documents were occurred in one day. As our data usually describes occurrence time in day granularity and the sample size is too small, we could not compare the features in finer granularity.

4 Conclusion and Discussion

In this paper, we tried to establish a methodological approach to extract distinctive features of a novel pandemic on the early symptoms experienced by the patients. We applied popular data mining techniques to swine flu suspected cases. The results correspond to outputs of previous specialized medical domain research. We could get valuable information with extremely small amount of data. This methodological approach could be used usefully in novel infectious disease management and research to prevent spreading of the pandemic at the very beginning stage.

Acknowledgement

This work was supported by the Korea Science and Engineering Foundation (**KOSEF**) grant funded by the Korea government(**MEST**) (No. 2009-0090853).

References

Kristine A. Himmerick. *H1N1 in perspective: The clinical impact of a novel influenza A virus.* JAAPA CME articles. December 01, 2009

Towards Event Extraction from Full Texts on Infectious Diseases

Sampo Pyysalo* **Tomoko Ohta*** **Han-Cheol Cho*** **Dan Sullivan†**
Chunhong Mao† **Bruno Sobral†** **Jun'ichi Tsujii*‡§** **Sophia Ananiadou‡§**
*Department of Computer Science, University of Tokyo, Tokyo, Japan
†Virginia Bioinformatics Institute, Virginia Tech, Blacksburg, Virginia, USA
‡School of Computer Science, University of Manchester, Manchester, UK
§National Centre for Text Mining, University of Manchester, Manchester, UK
{smp,okap,priancho,tsujii}@is.s.u-tokyo.ac.jp
{dsulliva,cmao,sobral}@vbi.vt.edu
Sophia.Ananiadou@manchester.ac.uk

Abstract

Event extraction approaches based on expressive structured representations of extracted information have been a significant focus of research in recent biomedical natural language processing studies. However, event extraction efforts have so far been limited to publication abstracts, with most studies further considering only the specific transcription factor-related subdomain of molecular biology of the GENIA corpus. To establish the broader relevance of the event extraction approach and proposed methods, it is necessary to expand on these constraints. In this study, we propose an adaptation of the event extraction approach to a subdomain related to infectious diseases and present analysis and initial experiments on the feasibility of event extraction from domain full text publications.

1 Introduction

For most of the previous decade, biomedical Information Extraction (IE) efforts have focused primarily on tasks that allow extracted information to be represented as simple pairs of related entities. This representation is applicable to many IE targets of interest, such as gene-disease associations (Chun et al., 2006) and protein-protein interactions (Nédellec, 2005; Krallinger et al., 2007). However, it has limited applicability to advanced applications such as semantic search, Gene Ontology term annotation, and pathway extraction, tasks for which and relatively few resources or systems (e.g. (Rzhetsky et al., 2004)) have been introduced. A number of recent studies have proposed more expressive representations of extracted information, introducing resources supporting advanced IE approaches (Pyysalo et al., 2007; Kim et al., 2008; Thompson et al., 2009; Ananiadou et al., 2010a). A significant step in the development of domain IE methods capable of extracting this class of representations was taken in the BioNLP'09 shared task on event extraction, where 24 teams participated in an IE task setting requiring the extraction of structured representations of multi-participant biological events of several types (Kim et al., 2009).

While the introduction of structured event extraction resources and methods has notably advanced the state of the art in biomedical IE representations, the focus of event extraction studies carries other limitations frequently encountered in domain IE efforts. Specifically, resources annotated for biomedical events contain exclusively texts from publication abstracts, typically further drawn from small subdomains of molecular biology. These choices constrain not only the types of texts but also the types of events considered, restricting the applicability of event extraction. This paper presents results from one ongoing effort to extend an event extraction approach over these boundaries, toward event extraction from full text documents in the domain of infectious diseases.

In this study, we consider the subdomain related to Type IV secretion systems as a model subdomain of interest within the broad infectious diseases domain. Type IV secretion systems (T4SS) are mechanisms for transferring DNA and proteins across cellular boundaries. T4SS are found in a broad range of Bacteria and in some Archaea. These translocation systems enable gene transfer across cellular membranes thus contributing to the spread of antibiotic resistance and viru-

Proceedings of the 2010 Workshop on Biomedical Natural Language Processing, ACL 2010, pages 132–140,
Uppsala, Sweden, 15 July 2010.

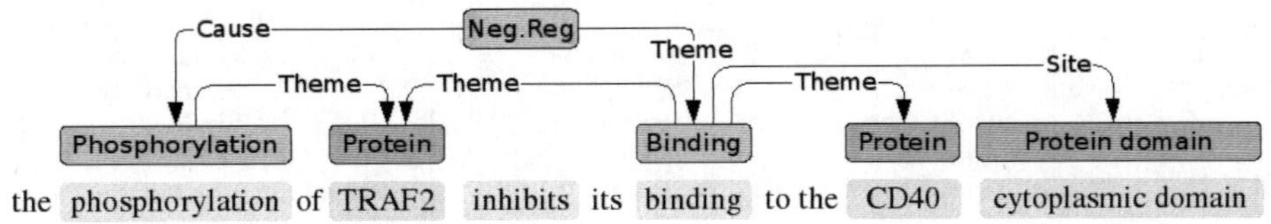

Figure 1: Event representation example. Inhibition of binding caused by phosphorylation is represented using three events. The shaded text background identifies the text bindings of the events and entities.

lence genes making them an especially important mechanism in infectious disease research (Juhas et al., 2008). Type IV secretion systems are found in plant pathogens, such as Agrobacterium tumefaciens, the cause of crown gall disease as well as in animal pathogens, such as Helicobacter pylori, a cause of severe gastric disease. The study of T4SS has been hampered by the lack of consistent terminology to describe genes and proteins associated with the translocation mechanism thus motivating the use of natural language processing techniques to enhance information retrieval and information extraction from relevant literature.

2 Event Extraction for the T4SS Domain

This section presents the application of an event extraction approach to the T4SS domain.

2.1 Event Extraction

We base our information extraction approach on the model introduced in the BioNLP'09 shared task on event extraction. Central to this approach is the event representation, which can capture the association of multiple participants in varying roles and numbers and treats events as primary objects of annotation, thus allowing events to be participants in other events. Further, both entities and events are text-bound, i.e. anchored to specific expressions in text (Figure 1).

The BioNLP'09 shared task defined nine event types and five argument types (roles): *Theme* specifies the core participant(s) that an event affects, *Cause* the cause of the the event, *Site* a specific domain or region on a participant involved in the event, and *ToLoc* and *AtLoc* locations associated with localization events (Table 1). Theme and Cause arguments may refer to either events or gene/gene product entities, and other arguments refer to other physical entities. The Theme argument is always mandatory, while others can be omitted when a relevant participant is not stated.

The event types were originally defined to capture statements of biologically relevant changes in the state of entities in a target subdomain involving transcription factors in human blood cells. In adapting the approach to new domains, some extension of the event types is expected to be necessary. By contrast, the argument types and the general design of the representation are intended to be general, and to maintain compatibility with existing systems we aim to avoid modifying these.

Event type	Args	Example
Gene expression	T	5-LOX is *coexpressed*
Transcription	T	IL-4 *transcription*
Protein catabolism	T	IkB-A *proteolysis*
Localization	T,L	*translocation* of STAT6
Phosphorylation	T,S	NF90 was *phosphorylated*
Binding	T+,S+	Nmi *interacts* with STAT
Regulation	T,C,S	IL-4 gene *control*
Positive regulation	T,C,S	IL-12 *induced* binding
Negative regulation	T,C,S	*suppressed* dimerization

Table 1: Event types targeted in the BioNLP'09 shared task and their arguments, with minimal examples of each event type. Arguments abbreviate for (T)heme, (C)ause, (S)ite and L for ToLoc/AtLoc, with "+" identifying arguments than can occur multiple times. The expression marked as triggering the event shown in italics.

2.2 T4SS Domain

A corpus of full-text publications relating to the T4SS subdomain of the infectious diseases domain annotated for biological entities and terms of interest to domain experts was recently introduced by (Ananiadou et al., 2010b). In the present study, we use this corpus as a reference standard defining domain information needs. In the following we briefly describe the corpus annotation and the view it provides of the domain.

The T4SS corpus annotation covers four classes of tagged entities and terms: Bacteria, Cellular components, Biological Processes, and Molecular functions. The latter three correspond to the three Gene Ontology (GO) (Ashburner et al., 2000) top-level sub-ontologies, and terms of these types were annotated with reference to both GO and relevance to the interests of domain experts, with guidelines

Bacterium		**Cell component**		**Biological process**		**Molecular function**	
A. tumefaciens	32.7%	T4SS	5.2%	virulence	14.1%	nucleotide-binding	20.3%
H. pylori	20.0%	Ti plasmid	5.1%	conjugation	7.9%	ATPase activity	17.3%
L. pneumophila	16.3%	outer membrane	4.2%	localization	6.1%	NTP-binding	14.7%
E. coli	12.3%	membrane	3.5%	nuclear import	5.8%	ATP-binding	12.2%
B. pertussis	3.0%	genome	3.4%	transfer	5.1%	DNA-binding	9.1%

Table 2: Most frequently tagged terms (after normalization) and their relative frequencies of all tagged entities of each of the four types annotated in the T4SS corpus.

Type	**Annotations**
Bacteria	529
Cellular component	2237
Biological process	1873
Molecular function	197

Table 3: Statistics for the existing T4SS corpus annotation.

requiring that marked terms be both found in GO and associated with T4SS. These constraints assure that the corpus is relevant to the information needs of biologists working in the domain and that it can be used as a reference for the study of automatic GO annotation. In the work introducing the corpus, the task of automatic GO annotation was studied as facilitating improved information access, such as advanced search functionality: GO annotation can allow for search by semantic classes or co-occurrences of terms of specified classes. The event approach considered in this study further extends on these opportunities in introducing a model allowing e.g. search by specific associations of the concepts of interest.

The previously created annotation of the T4SS corpus covers 27 full text publications totaling 15143 pseudo-sentences (text sentences plus table rows, references, etc.) and 244942 tokens.[1] A total of nearly 5000 entities and terms are annotated in these documents; Table 2 shows the most frequently tagged terms of each type after basic normalization of different surface forms, and Table 3 gives the per-class statistics. Domain characteristics are clearly identifiable in the first three tagged types, showing disease-related bacteria, their major cellular components, and processes related to movement, reproduction and infection. The last term type is dominated by somewhat more generic binding-type molecular functions.

In addition to the four annotated types it was recognized during the original T4SS corpus annotation that genes and gene products are centrally important for domain information needs, but their annotation was deferred to focus on novel categories. As part of the present study, we introduce annotation for gene/gene product (GGP) mentions (Section 3.2), and in the following discussion of applying an event extraction approach to the domain the availability of this class annotation as an additional category is assumed.

2.3 Adaptation of the Event Model

The event model involves two primary categories of representation: physical entities such as genes and proteins are elementary (non-structured) and their mentions annotated as typed spans of text,[2] and events and processes ("things that happen") are represented using the structured event representation described in Section 2.1. This division applies straightforwardly to the T4SS annotations, suggesting an approach where bacteria and cell components retain their simple tagged-term representation and the biological processes and molecular functions are given an event representation. In the following, we first analyze correspondences between the latter two classes and BioNLP'09 shared task events, and then proceed to study the event arguments and their roles as steps toward a complete event model for the domain.

Molecular functions, the smallest class tagged in the T4SS corpus, are highly uniform: almost 75% involve binding, immediately suggesting representation using the Binding class of events defined in the applied event extraction model. The remaining functions are *ATPase activity*, together with its exact GO synonyms (e.g. *ATP hydrolase activity*) accounting for 19% of the terms, the general type *hydrolysis* (4.5%), and a small number of rare other functions. While these have no correspondence with previously defined event types,

[1]While the document count is modest compared to that of abstract-based corpora, we estimate that in terms of the amount of text (tokens) the corpus corresponds to over 1000 abstracts, comparable in size to e.g. the GENIA event corpus (Kim et al., 2008).

[2]Normalization identifying e.g. the Uniprot entry corresponding to a protein mention may also be necessary, but here excluded from consideration an independent issue.

Class	Category	Freq
	Transfer	27.6%
Location	Localization	15.6%
	Import/export	14.5%
	Virulence	14.1%
High-level	Assembly	8.7%
process	Conjugation	8.3%
	Secretion	8.1%
(Other)		1.8%

Table 4: Categorization of T4SS corpus biological processes and relative frequency of mentions of each category of the total tagged.

their low overall occurrence counts make them of secondary interest as extraction targets.

The biological processes are considerably more diverse. To identify general categories, we performed a manual analysis of the 217 unique normalized terms annotated in the corpus as biological processes (Table 4). We find that the majority of the instances (58%) relate to location or movement. As related types of statements are annotated as Localization events in the applied model, we propose to apply this event type and differentiate between the specific subtypes on the basis of the event arguments. A further 39% are of categories that can be viewed as high-level processes. These are distinct from the events considered in the BioNLP'09 shared task in involving coarser-grained events and larger-scale participants than the GGP entities considered in the task: for example, conjugation occurs between bacteria, and virulence may involve a human host.

To analyze the role types and arguments characteristic of domain events, we annotated a small sample of tagged mentions for the most frequent types in the broad classification discussed above: Binding for Molecular function, Transfer for Location-related, and Virulence for High-level process. The statistics of the annotated 65 events are shown in Tables 5, 6 and 7. For Binding, we find that while an estimated 90% of events involve a GGP argument, the other participant of the binding is in all cases non-GGP, most frequently of Nucleotide type (e.g. NTP/ATP). While only GGP Binding arguments were considered in the shared task events, the argument structures are typical of multi-participant binding and this class of expressions are in scope of the original GENIA Event corpus annotation (Kim et al., 2008). Event annotations could thus potentially be derived from existing data. Localization event arguments show substantially greater variety and

Freq	Arguments
78%	Theme: GGP, Theme: Nucleotide
5.5%	Theme: GGP, Theme: DNA
5.5%	Theme: GGP, Theme: Sugar
5.5%	Theme: Protein family, Theme: DNA
5.5%	Theme: Protein, Theme: Nucleotide

Table 5: Binding event arguments.

Freq	Arguments
16%	Theme: DNA, From/To: Organism
16%	Theme: DNA
16%	Theme: Cell component
12%	Theme: DNA, To: Organism
8%	Theme: Protein family, From/To: Organism
4%	Theme: GGP
4%	Theme: GGP, To: Organism
4%	Theme: GGP, From: Organism
4%	Theme: Protein family, From: Organism
4%	Theme: Protein family
4%	Theme: Organism, To: Cell component
4%	Theme: DNA From: Organism, To: Cell component
4%	(no arguments)

Table 6: Localization (Transfer) event arguments.

Freq	Arguments
64%	Cause: GGP
16%	Theme:Organism, Cause: GGP
8%	Cause: Organism
8%	(no arguments)
4%	Cause: Protein family

Table 7: Process (Virulence) arguments.

some highly domain-specific argument combinations, largely focusing on DNA and Cell component (e.g. phagosome) transfer, frequently involving transfer between different organisms. While the participants are almost exclusively of types that do not appear in Localization events in existing annotations, the argument structures are standard and in our judgment reasonably capture the analyzed statements, supporting the applicability of the general approach. Finally, the argument analysis shown in Table 7 supports the previous tentative observation that the high-level biological processes are notably different from previously considered event types: for over 80% of these processes no overtly stated *Theme* could be identified. We take this to indicate that the themes – the core participants that the processes concern – are obvious in the discourse context and their overt expression would be redundant. (For example, in the context *virulence* obviously involves a host and *conjugation* involves bacteria.) By contrast, in the corpus the entities contributing to these processes are focused: a participant we have here analyzed as *Cause* is stated in over 90% of cases. This

	Sentences	Tokens
Abstracts	150	3789
Full texts	448	13375
Total	598	17164

Table 8: Statistics for the selected subcorpus.

	GGP	GGP/sentence
Abstracts	124	0.82
Full texts	394	0.88
Total	518	0.87

Table 9: Statistics for the GGP annotation.

novel pattern of event arguments suggests that the event model should be augmented to capture this category of high-level biological processes. Here, we propose an event representation for these processes that removes the requirement for a Theme and substitutes instead a mandatory Cause as the core argument. In the event annotation and experiments, we focus on this newly proposed class.

3 Annotation

This section describes the new annotation introduced for the T4SS corpus.

3.1 Text Selection

The creation of exhaustive manual annotation for the full T4SS corpus represents a considerable annotation effort. Due to resource limitations, for this study we did not attempt full-scope annotation but instead selected a representative subset of the corpus texts. We aimed to select texts that provide good coverage of the text variety in the T4SS corpus and can be freely redistributed for use in research. We first selected for annotation all corpus documents with at least a freely available PubMed abstract, excluding 3 documents. As the corpus only included a single freely redistributable Open Access paper, we extended full text selection to manuscripts freely available as XML/HTML (i.e. not only PDF) via PubMed Central. While these documents cannot be redistributed in full, their text can be reliably combined with standoff annotations to recreate the annotated corpus.

In selected full-text documents, to focus annotation efforts on sections most likely to contain reliable new information accessible to natural language processing methods, we further selected the publication body text, excluding figures and tables and their captions, and removed Methods and Discussion sections. We then removed artifacts such as page numbers and running heads and cleaned remaining errors from PDF conversion of the original documents. This selection produced a subcorpus of four full-text documents and 19 abstracts. The statistics for this corpus are shown in Table 8.

3.2 Gene/Gene Product Annotation

As gene and gene product entities are central to domain information needs and the core entities of the applied event extraction approach, we first introduced annotation for this entity class. We created manual GGP annotation following the annotation guidelines of the GENIA GGP Corpus (Ohta et al., 2009). As this corpus was the source of the gene/protein entity annotation provided as the basis of the BioNLP shared task on event extraction, adopting its annotation criteria assures compatibility with recently introduced event extraction methods. Briefly, the guidelines specify tagging for minimal continuous spans of specific gene/gene product names, without differentiating between DNA/RNA/protein. A "specific name" is understood to be a a name that allows a domain expert to identify the entry in a relevant database (Entrez gene/Uniprot) that the name refers to. Only GGP names are tagged, excluding descriptive references and the names of related entities such as complexes, families and domains.

The annotation was created on the basis of an initial tagging created by augmenting the output of the BANNER tagger (Leaman and Gonzalez, 2008) by dictionary- and regular expression-based tagging. This initial high-recall markup was then corrected by a human annotator. To confirm that the annotator had correctly identified subdomain GGPs and to check against possible error introduced through the machine-assisted tagging, we performed a further verification of the annotation on approx. 50% of the corpus sentences: we combined the machine- and human-tagged annotations as candidates, removed identifying information, and asked two domain experts to identify the correct GGPs. The two sets of independently produced judgments showed very high agreement: holding one set of judgments as the reference standard, the other would achieve an f-score of 97% under the criteria presented in Section 4.2. We note as one contributing factor to the high agreement that the domain has stable and systematically applied GGP naming criteria. The statistics of the full GGP annotation are shown in Table 9.

	Events	Event/sentence
Abstracts	15	0.1
Full texts	5	0.01
Additional	80	2.2
Total	100	0.16

Table 10: Statistics for the event annotation.

3.3 Event Annotation

Motivated by the analysis described in Section 2.3, we chose to focus on the novel category of associations of GGP entities in high-level processes. Specifically, we chose to study biological processes related to virulence, as these are the most frequent case in the corpus and prototypical of the domain. We adopted the GENIA Event corpus annotation guidelines (Kim et al., 2008), marking associations between specific GGPs and biological processes discussed in the text even when these are stated speculatively or their existence explicitly denied. As the analysis indicated this category of processes to typically involve a single stated participant in a fixed role, annotations were initially recorded as (GGP, process) pairs and later converted into an event representation.

During annotation, the number of annotated GGP associations with the targeted class of processes in the T4SS subcorpus was found to be too low to provide material for both training and testing a supervised learning-based event extraction approach. To extend the source data, we searched PubMed for cases where a known T4SS-related protein co-occurred with an expression known to relate to the targeted process class (e.g. *virulence*, *virulent*, *avirulent*, *non-virulent*) and annotated a further set of sentences from the search results for both GGPs and their process associations. As the properties of these additional examples could not be assured to correspond to those of the targeted domain texts, we used these annotations only as development and training data, performing evaluation on cases drawn from the T4SS subcorpus.

As the annotation target was novel, we performed two independent sets of judgments for all annotated cases, jointly resolving disagreements. Although initial agreement was low, for a final set of judgments we measured high agreement, corresponding to 93% f-score when holding one set of judgments as the gold standard. The statistics of the annotation are shown in Table 10. Annotations are sparse in the T4SS subcorpus and, as expected, very dense in the targeted additional data.

4 Experiments

4.1 Methods

For GGP tagging experiments, we applied a state-of-the-art tagger with default settings as reference and a custom tagger for adaptation experiments. As the reference tagger, we applied a recent release of BANNER (Leaman and Gonzalez, 2008) trained on the GENETAG corpus (Tanabe et al., 2005). The corpus is tagged for gene and protein-related entities and its texts drawn from a broad selection of PubMed abstracts. The current revision of the tagger[3] achieves an f-score of 86.4% on the corpus, competitive with the best result reported in the BioCreative II evaluation (Wilbur et al., 2007), 87.2%. The custom tagger[4] follows the design of BANNER in both the choice of Conditional Random Fields (Lafferty et al., 2001) as the applied learning method and the basic feature design, but as a key extension can further adopt features from external dictionaries as both positive and negative indicators of tagged entities. Tagging experiments were performed using a document-level 50/50 split of the GGP-annotated subcorpus.

For event extraction, we applied an adaptation of the approach of the top-ranking system in the BioNLP'09 shared task (Björne et al., 2009): all sentences in the input text were parsed with the McClosky-Charniak (2008) parser and the resulting phrase structure analyses then converted into the Stanford Dependency representation using conversion included in the Stanford NLP tools (de Marneffe et al., 2006). Trigger recognition was performed with a simple regular expression-based tagger covering standard surface form variation. Edge detection was performed using a supervised machine learning approach, applying the LibSVM (Chang and Lin, 2001) Support Vector Machine implementation with a linear kernel and the feature representation of Björne et al˙ (2009), building largely around the shortest dependency path connecting a detected trigger with a candidate participant. The SVM regularization parameter was selected by a sparse search of the parameter space with evaluation using cross-validation on the training set. As the class of events targeted for extraction in this study are of a highly restricted type, each taking only of a single mandatory Cause argument, the construction of events from detected

[3]http://banner.sourceforge.net

[4]http://www-tsujii.is.s.u-tokyo.ac.jp/NERsuite/

	Precision	Recall	F-score
Abstracts	68.1%	89.5%	77.3%
Full texts	56.9%	80.7%	66.7%
Total	59.4%	82.8%	69.2%

Table 11: Initial GGP tagging results.

triggers and edges could be implemented as a simple deterministic rule.

4.2 Evaluation Criteria

For evaluating the performance of the taggers we apply a relaxed matching criterion that accepts a match between an automatically tagged and a gold standard entity if the two overlap at least in part. This relaxation is adopted to focus on true tagging errors. The GENETAG entity span guidelines differ from the GENIA GGP guidelines adopted here in allowing the inclusion of e.g. head nouns when names appear in modifier position, while the annotation guidelines applied here require marking only the minimal name.[5] When applying strict matching criteria, a substantial number of errors may trace back to minor boundary differences (Wang et al., 2009), which we consider of secondary interest to spurious or missing tags. Overall results are microaverages, that is, precision, recall and f-score are calculated from the sum of true positive etc. counts over individual documents.

For event extraction, we applied the BioNLP'09 shared task event extraction criteria (Kim et al., 2009) with one key change: to make it possible to evaluate the extraction of the high-level process participants, we removed the requirement that all events must define a Theme as their core argument.

4.3 Gene/Gene Product Tagging

The initial GGP tagging results using BANNER are shown in Table 11. We find that even for the relaxed overlap matching criterion, the f-score is nearly 10% points lower than reported on GENETAG in the evaluation on abstracts. For full texts, performance is lower yet by a further 10% points. In both cases, the primary problem is the poor precision of the tagger, indicating that many non-GGPs are spuriously tagged.

To determine common sources of error, we performed a manual analysis of 100 randomly selected falsely tagged strings (Table 12). We find that the most frequent category consists of cases that are arguably correct by GENETAG annotation criteria, which allow named protein families of groups to be tagged. A similar argument can be made for domains or regions. Perhaps not surprisingly, a large number of false positives relate to features common in full texts but missing from the abstracts on which the tagger was trained, such as figure and table references. Finally, systematic errors are made for entities belonging to other categories such as named cell components or species.

Category	Freq	Examples
GGP family or group	34%	VirB, tmRNA genes
Figure/table	26%	Fig. 1B, Table 1
Cell component	10%	T4SS, ER vacuole
Species/strain	9%	E. coli, A348deltaB4.5
Misc.	9%	step D, Protocol S1
GGP domain or region	4%	Pfam domain
(Other)	8%	TrIP, LGT

Table 12: Common sources of false positives in GGP tagging.

	Precision	Recall	F-score
Abstracts	90.5%	95.7%	93.1%
Full texts	90.0%	93.2%	91.6%
Total	90.1%	93.8%	91.9%

Table 13: GGP tagging results with domain adaptation.

To address these issues, we applied a domain-adapted custom tagger that largely replicates the features of BANNER, further integrating information from the UMLS Metathesaurus,[6] which provides a large dictionary containing terms covering 135 different semantic classes, and a custom dictionary of 1081 domain GGP names, compiled by (Ananiadou et al., 2010b). The non-GGP UMLS Metathesaurus terms provided negative indicators for reducing spurious taggings, and the custom dictionary positive indicators. Finally, we augmented the GENETAG training data with 10 copies[7] of the training half of the T4SS GGP corpus as in-domain training data.

Table 13 shows the results with the domain-adapted tagger. We find dramatically improved performance for both abstracts and full texts, showing results competitive with the state of the art performance on GENETAG (Wilbur et al., 2007). Thus, while the performance of an unadapted tagger falls short of both results reported

[5] GENETAG annotations include e.g. *human ets-1 protein*, whereas the guidelines applied here would require marking only *ets-1*.

[6] `http://www.nlm.nih.gov/research/umls/`

[7] As the GENETAG corpus is considerably larger than the T4SS GGP corpus, replication was used to assure that sufficient weight is given to the in-domain data in training.

	Precision	Recall	F-score
Co-occurrence	65%	100%	78%
Machine learning	81%	85%	83%

Table 14: Event extraction results.

on GENETAG and levels necessary for practical application, adaptation addressing common sources of error through the adoption of general and custom dictionaries and the use of a small set of in-domain training data was successful in addressing these issues. The performance of the adapted tagger is notably high given the modest size of the in-domain data, perhaps again reflecting the consistent GGP naming conventions of the subdomain.

4.4 Event Extraction

We performed an event extraction experiment following the training and test split described in Section 3.3. Table 14 shows the results of the applied machine learning-based method contrasted with a co-occurrence baseline replacing the edge detection with a rule that extracts a Cause edge for all trigger-GGP combinations co-occurring within sentence scope. This approach achieves 100% recall as the test data was found to only contain events where the arguments are stated in the same sentence as the trigger.

The results show that the machine learning approach achieves very high performance, matching the best results reported for any single event type in the BioNLP'09 shared task (Kim et al., 2009). The very high co-occurrence baseline result suggests that the high performance largely reflects the relative simplicity of the task. With respect to the baseline result, the machine-learning approach achieves a 21% relative reduction in error.

While this experiment is limited in both scope and scale, it suggests that the event extraction approach can be beneficially applied to detect domain events represented by novel argument structures. As a demonstration of feasibility the result is encouraging for both the applicability of event extraction to this specific new domain and for the adaptability of the approach to new domains in general.

5 Discussion and Conclusions

We have presented a study of the adaptation of an event extraction approach to the T4SS subdomain as a step toward the introduction of event extraction to the broader infectious diseases domain. We applied a previously introduced corpus of subdomain full texts annotated for mentions of bacteria and terms from the three top-level Gene Ontology subontologies as a reference defining domain information needs to study how these can be met through the application of events defined in the BioNLP'09 Shared Task on event extraction. Analysis indicated that with minor revision of the arguments, the Binding and Localization event types could account for the majority of both biological processes and molecular functions of interest. We further identified a category of "high-level" biological processes such as the *virulence* process typical of the subdomain, which necessitated extension of the considered event extraction model.

Based on argument analysis, we proposed a representation for high-level processes in the event model that substitutes Cause for Theme as the core argument. We further produced annotation allowing an experiment on the extraction of the dominant category of virulence processes with gene/gene product (GGP) causes, annotating 518 GGP mentions and 100 associations between these and the processes. Experiments indicated that with annotated in-domain resources both the GGP entities and their associations with processes could be extracted with high reliability.

In future work we will extend the model and annotation proposed in this paper to the broader infectious diseases domain, introducing annotated resources and extraction methods for advanced information access. All annotated resources introduced in this study are available from the GENIA project homepage.[8]

Acknowledgments

This work was partially supported by Grant-in-Aid for Specially Promoted Research (MEXT, Japan), the National Institutes of Health, grant number HHSN272200900040C, and the Joint Information Systems Committee (JISC, UK).

References

Sophia Ananiadou, Sampo Pyysalo, Junichi Tsujii, and Douglas B. Kell. 2010a. Event extraction for systems biology by text mining the literature. *Trends in Biotechnology*. (to appear).

[8] http://www-tsujii.is.s.u-tokyo.ac.jp/GENIA/

Sophia Ananiadou, Dan Sullivan, Gina-Anne Levow, Joseph Gillespie, Chunhong Mao, Sampo Pyysalo, Jun'ichi Tsujii, and Bruno Sobral. 2010b. Named entity recognition for bacterial type IV secretion systems. (manuscript in review).

M Ashburner, CA Ball, JA Blake, D Botstein, H Butler, JM Cherry, AP Davis, K Dolinski, SS Dwight, JT Eppig, MA Harris, DP Hill, L Issel-Tarver, A Kasarskis, S Lewis, JC Matese, JE Richardson, M Ringwald, GM Rubin, and G Sherlock. 2000. Gene ontology: tool for the unification of biology. *Nature genetics*, 25:25–29.

Jari Björne, Juho Heimonen, Filip Ginter, Antti Airola, Tapio Pahikkala, and Tapio Salakoski. 2009. Extracting complex biological events with rich graph-based feature sets. In *Proceedings of the BioNLP 2009 Workshop Companion Volume for Shared Task*, pages 10–18, Boulder, Colorado, June. Association for Computational Linguistics.

Chih-Chung Chang and Chih-Jen Lin, 2001. *LIBSVM: a library for support vector machines*. Software available at `http://www.csie.ntu.edu.tw/~cjlin/libsvm`.

Hong-Woo Chun, Yoshimasa Tsuruoka, Jin-Dong Kim, Rie Shiba, Naoki Nagata, Teruyoshi Hishiki, and Jun'ichi Tsujii. 2006. Extraction of gene-disease relations from medline using domain dictionaries and machine learning. In *Proceedings of the Pacific Symposium on Biocomputing (PSB'06)*, pages 4–15.

Marie-Catherine de Marneffe, Bill MacCartney, and Christopher D. Manning. 2006. Generating Typed Dependency Parses from Phrase Structure Parses. In *Proceedings of the Fifth International Conference on Language Resources and Evaluation (LREC'06)*, pages 449–454.

Mario Juhas, Derrick W. Crook, and Derek W. Hood. 2008. Type IV secretion systems: tools of bacterial horizontal gene transfer and virulence. *Cellular microbiology*, 10(12):2377–2386.

Jin-Dong Kim, Tomoko Ohta, and Jun'ichi Tsujii. 2008. Corpus annotation for mining biomedical events from lterature. *BMC Bioinformatics*, 9(1):10.

Jin-Dong Kim, Tomoko Ohta, Sampo Pyysalo, Yoshinobu Kano, and Jun'ichi Tsujii. 2009. Overview of BioNLP'09 Shared Task on Event Extraction. In *Proceedings of Natural Language Processing in Biomedicine (BioNLP) NAACL 2009 Workshop*, pages 1–9.

Martin Krallinger, Florian Leitner, and Alfonso Valencia. 2007. Assessment of the Second BioCreative PPI task: Automatic Extraction of Protein-Protein Interactions. In L. Hirschman, M. Krallinger, and A. Valencia, editors, *Proceedings of Second BioCreative Challenge Evaluation Workshop*, pages 29–39.

John D. Lafferty, Andrew McCallum, and Fernando C . N. Pereira. 2001. Conditional random fields: Probabilistic models for segmenting and labeling sequence data. In *ICML '01: Proceedings of the 18th International Conference on Machine Learning*, pages 282–289.

R. Leaman and G. Gonzalez. 2008. Banner: an executable survey of advances in biomedical named entity recognition. *Pacific Symposium on Biocomputing*, pages 652–663.

David McClosky and Eugene Charniak. 2008. Self-Training for Biomedical Parsing. In *Proceedings of the 46th Annual Meeting of the Association for Computational Linguistics - Human Language Technologies (ACL-HLT'08)*, pages 101–104.

Claire Nédellec. 2005. Learning Language in Logic - Genic Interaction Extraction Challenge. In J. Cussens and C. Nédellec, editors, *Proceedings of the 4th Learning Language in Logic Workshop (LLL05)*, pages 31–37.

Tomoko Ohta, Jin-Dong Kim, Sampo Pyysalo, and Jun'ichi Tsujii. 2009. Incorporating GENETAG-style annotation to GENIA corpus. In *Proceedings of Natural Language Processing in Biomedicine (BioNLP) NAACL 2009 Workshop*, pages 106–107, Boulder, Colorado. Association for Computational Linguistics.

Sampo Pyysalo, Filip Ginter, Juho Heimonen, Jari Björne, Jorma Boberg, Jouni Järvinen, and Tapio Salakoski. 2007. BioInfer: A corpus for information extraction in the biomedical domain. *BMC Bioinformatics*, 8(50).

Andrey Rzhetsky, Ivan Iossifov, Tomohiro Koike, Michael Krauthammer, Pauline Kra, Mitzi Morris, Hong Yu, Pablo Ariel Duboué, Wubin Weng, W. John Wilbur, Vasileios Hatzivassiloglou, and Carol Friedman. 2004. GeneWays: A system for extracting, analyzing, visualizing, and integrating molecular pathway data. *Journal of Biomedical Informatics*, 37(1):43–53.

Lorraine Tanabe, Natalie Xie, Lynne Thom, Wayne Matten, and John Wilbur. 2005. Genetag: a tagged corpus for gene/protein named entity recognition. *BMC Bioinformatics*, 6(Suppl 1):S3.

Paul Thompson, Syed Iqbal, John McNaught, and Sophia Ananiadou. 2009. Construction of an annotated corpus to support biomedical information extraction. *BMC Bioinformatics*, 10(1):349.

Yue Wang, Jin-Dong Kim, Rune Saetre, Sampo Pyysalo, and Jun'ichi Tsujii. 2009. Investigating heterogeneous protein annotations toward cross-corpora utilization. *BMC Bioinformatics*, 10(1):403.

John Wilbur, Lawrence Smith, and Lorraine Tanabe. 2007. BioCreative 2. Gene Mention Task. In L. Hirschman, M. Krallinger, and A. Valencia, editors, *Proceedings of Second BioCreative Challenge Evaluation Workshop*, pages 7–16.

Applying the TARSQI Toolkit to augment text mining of EHRs

Amber Stubbs
Department of Computer Science
Brandeis University MS 018
Waltham, Massachusetts, 02454 USA
astubbs@cs.brandeis.edu

Benjamin Harshfield
Channing Laboratory
Brigham and Women's Hospital
Boston, Massachusetts, 02115 USA
rebjh@channing.harvard.edu

Abstract

We present a preliminary attempt to apply the TARSQI Toolkit to the medical domain, specifically electronic health records, for use in answering temporally motivated questions.

1 Introduction

Electronic Health Records are often the most complete records of a patient's hospital stay, making them invaluable for retrospective cohort studies. However, the free text nature of these documents makes it difficult to extract complex information such as the relative timing of conditions or procedures. While there have been recent successes in this endeavor (Irvine et al., 2008; Mowery et al., 2009; Zhou et al., 2007), there is still much to be done. We present work done to adapt the TARSQI Toolkit (TTK) to the medical domain. Though the use of the TTK and a set of auxiliary Perl scripts, we perform information extraction over a set of 354 discharge summaries used in the R3i REALIST study to answer the following question:

> Which patients can be positively identified as being on statins at the time they were admitted to the hospital?

2 TARSQI Toolkit

The TARSQI Toolkit, developed as a part of the AQUAINT workshops, is a "modular system for automatic temporal and event annotation of natural language" in newswire texts (Verhagen and Pustejovsky, 2008). The different modules preprocess the data, label events and times, create links between times and events (called "tlinks"), and mark subordination relationships. Output from the TTK consists documents annotated in TimeML, an XML specification for event and time annotation (Pustejovsky et al., 2005). Of particular interest for this project are EVITA, the module responsible for finding events in text, and Blinker, the module used to create syntactic rule-based links between events and timexes.

3 Structure of EHRs

The bodies of the Electronic Health Records used were segmented, with each section having a header indicating the topic of that section ("Medical History", "Course of Treatment", "Discharge Medications", etc). Header names and sections are not standardized across EHRs, but often give important temporal information about when events described took place (Denny et al., 2008).

4 Statin Extraction Methodology

As the purpose of this task was to discover what changes to the TTK would be necessary to make the transition from newswire to medical texts, over the course of two weeks we filled in the gaps in the toolkit's abilities with a few auxiliary Perl scripts. Specifically, these scripts were used to clean up input so that it conformed to TTK expectations, label the statins as events, locate section headers and associate temporal information with the headers.

A list of statins was acquired from an MD, and then supplemented with information from websites in order to get all currently marketed versions of the drugs. This list was then used in conjunction with a Perl script to find mentions of statins in the discharge summaries and create TimeML event tags for them.

In order to identify and categorize section headers we developed a program to automatically collect header names from a separate set of approximately 700 discharge summaries. Then we gathered statistics on word frequency and created simple rules for characterizing headers based on keywords. Headers were divided into four simple categories: Past, Present, After, and Not (for cate-

Proceedings of the 2010 Workshop on Biomedical Natural Language Processing, ACL 2010, pages 141–143,
Uppsala, Sweden, 15 July 2010.

gories that did not contain specific or relevant temporal information).

The Blinker component of the TTK was then modified to take into account temporal information stored in the header in addition to the syntactic information present in each individual sentence for the creation of tlinks.

5 Results

Output from the modified TTK was compared to the judgment of human annotators on the same dataset. Two annotators, employees of BWH/Harvard Medical involved in data management and review for clinical trials, were asked to label each file as yes for those patients taking statins at the time they were admitted to the hospital, and no for those that werent. Files where statins were mentioned without clear temporal anchorings were categorized as "unsure".

Inter-annotator agreement was 85% (Cohen kappa=.75), with 75% of the disagreements being between "no" and "unsure". The majority of these ambiguous cases were discharge summaries where a statin was listed under "discharge" but admission medications were not listed, nor were the statins mentioned as being started at the hospital. The annotation guidelines have been updated to reflect how to annotate these cases in the future. Overall, 139 patients were identified as being on statins, 174 were not on statins, and 41 were unclear.

As the question was which patients could be positively identified as being on statins at the time of admission, the files labeled as "unsure" were considered to be "no" for the purposes of evaluation against the TTK, making the totals 139 yeses to 215 noes. The comparison between human and computer annotation are shown below:

	Yes	**No**
Human	139	215
TTK	129	225

Table 1: Distribution of statin classifications.

The TTK system had an accuracy of 84% overall, with an accuracy of 95% on the files that the human annotators found to be unambiguous.

6 Limitations

While we were pleased by these results, a number of factors worked in the favor or the automated system. The task itself, while requiring a mixture of lexical and temporal knowledge, was greatly simplified by a finite list of medications and a binary outcome variable. Obscure abbreviations or misspellings could have prevented identification of statin mentions for both the computer and humans, making the overall accuracy questionable. Additionally, in the majority of documents the statins were mentioned in lists under temporally anchored headings rather than free text, thereby minimizing the impact of uncertain times as described in Hripcsak et al (2009).

7 Future work

Our work so far shows promising results for being able to modify the TARSQI Toolkit for use in the medical domain. In the future, we would like to integrate the functionality of the Perl scripts used in this project into the TTK, in particular expanding the vocabulary of the EVITA module to the medical domain, section header labeling, and the use of the headers in tlink creation.

New annotation schemas will need to be added to the project in order to get a more complete and accurate view of medical records. Under consideration is the Clinical E-Science Framework (CLEF) (Roberts et al., 2007) for annotating medical entities, actions (which would overlap with TimeML events), drugs, etc. Modifications to Blinker will be more fully integrated with the existing rule libraries. At this point it is unclear whether the TTK will remain a single program, or if it will split into domain-specific versions.

Furthermore, the number of files labeled "unsure" by human annotators highlights the need for cross-document analysis abilities. Had previous records for these patients been available, it seems likely that there would have been fewer uncertainties.

8 Conclusion

Modifying the TARSQI Toolkit, a newswire-trained parser, for application in the medical domain provided accurate results for a very specific time-sensitive query.

Acknowledgments

Partial support for the work described here was provided by the Residual Risk Reduction Initiative Foundation (r3i.org).

References

Joshua C Denny, Randolph A Miller, Kevin B Johnson, and Anderson Spickard. 2008. Development and evaluation of a clinical note section header terminology. *AMIA Annual Symposium proceedings*, pages 156–60.

George Hripcsak, Noémie Elhadad, Yueh-Hsia Chen, Li Zhou, and Frances P Morrison. 2009. Using empiric semantic correlation to interpret temporal assertions in clinical texts. *J Am Med Inform Assoc*, 16(2):220–7.

Ann K Irvine, Stephanie W Haas, and Tessa Sullivan. 2008. Tn-ties: A system for extracting temporal information from emergency department triage notes. *AMIA Annual Symposium proceedings*, pages 328–32.

Danielle L. Mowery, Henk Harkema, John N. Dowling, Jonathan L. Lustgarten, and Wendy W. Chapman. 2009. Distinguishing historical from current problems in clinical reports: which textual features help? In *BioNLP '09: Proceedings of the Workshop on BioNLP*, pages 10–18, Morristown, NJ, USA. Association for Computational Linguistics.

James Pustejovsky, Bob Ingria, and Roser Sauri et al., 2005. *The Language of Time: A Reader*, chapter The Specification Language TimeML, pages 545–558. Oxford University Press, Oxford.

Angus Roberts, Robert Gaizauskas, and Mark et al Hepple. 2007. The clef corpus: semantic annotation of clinical text. *AMIA Annual Symposium proceedings*, pages 625–9.

Marc Verhagen and James Pustejovsky. 2008. Temporal processing with the tarsqi toolkit. In *Coling 2008: Companion volume Posters and Demonstrations*, pages 189–192, Manchester, UK.

Li Zhou, Simon Parsons, and George Hripcsak. 2007. The evaluation of a temporal reasoning system in processing clinical discharge summaries. *J Am Med Inform Assoc*, 15(1):99–106.

Integration of Static Relations to Enhance Event Extraction from Text

Sofie Van Landeghem[1,2], **Sampo Pyysalo**[3], **Tomoko Ohta**[3], **Yves Van de Peer**[1,2]
1. Dept. of Plant Systems Biology, VIB, Gent, Belgium
2. Dept. of Plant Biotechnology and Genetics, Ghent University, Gent, Belgium
3. Department of Computer Science, University of Tokyo, Tokyo, Japan
yves.vandepeer@psb.vib-ugent.be

Abstract

As research on biomedical text mining is shifting focus from simple binary relations to more expressive event representations, extraction performance drops due to the increase in complexity. Recently introduced data sets specifically targeting static relations between named entities and domain terms have been suggested to enable a better representation of the biological processes underlying annotated events and opportunities for addressing their complexity. In this paper, we present the first study of integrating these static relations with event data with the aim of enhancing event extraction performance. While obtaining promising results, we will argue that an event extraction framework will benefit most from this new data when taking intrinsic differences between various event types into account.

1 Introduction

Recently, biomedical text mining tools have evolved from extracting simple binary relations between genes or proteins to a more expressive event representation (Kim et al., 2009). Furthermore, new data sets have been developed targeting relations between genes and gene products (GGPs) and a broader category of entities, covering terms that can not be annotated as named entities (NEs) but that are still highly relevant for biomedical information extraction (Ohta et al., 2009b). In contrast to relations involving change or causality, the annotation for this data covers relations such as *part-of*, here termed "static relations" (SR) (Pyysalo et al., 2009).

Tissue-specific expression of **interleukin-3**
expression event *GGP*

is mediated via cis-acting elements located
regulation event *term part-of GGP*

within 315 base pairs of the transcription start.
term part-of GGP

Figure 1: A sentence from PMID:8662845, showing how the event data set (single line) and the SR data set (double line) offer complementary information, enabling a more precise model of the biological reality.

As an example, Figure 1 depicts a sentence containing complementary annotations from the event data set and the SR data. The event annotation indicates an expression event involving the GGP "interleukin-3". Furthermore, regulation of this expression event is stated by the trigger word "mediated". In addition, the SR annotation marks two terms that refer to parts of the GGP, namely "cis-acting elements" and "transcription starts". These two terms provide more detailed information on the regulation event. Thus, by combining the two types of annotation, a text mining algorithm will be able to provide a more detailed representation of the extracted information. This would be in particular beneficial in practical applications such as abstract summarization or integration of the predictions into complex regulatory pathways.

In addition to providing enhanced representation of biological processes, the SR data set also offers interesting opportunities to improve on event extraction. As an example, consider the sentence presented in Figure 2, in which "c-Rel" and "p50" are both annotated as being subunits of the

Proceedings of the 2010 Workshop on Biomedical Natural Language Processing, ACL 2010, pages 144–152,
Uppsala, Sweden, 15 July 2010. ©2010 Association for Computational Linguistics

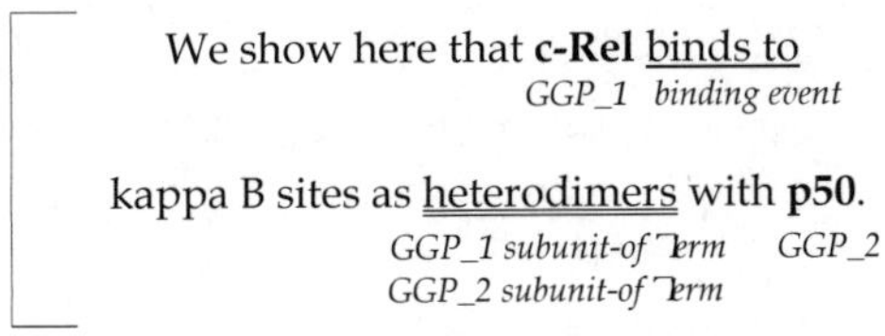

Figure 2: A sentence from PMID:1372388, showing how SR data (double line) can provide strong clues for the extraction of biomolecular events (double line) from text.

Event type	Args	Train	Devel	Test
Gene expression	T	1738	356	722
Transcription	T	576	82	137
Protein catabolism	T	110	21	14
Localization	T	265	53	174
Phosphorylation	T	169	47	139
Binding	T+	887	249	349
Regulation	T, C	961	173	292
Positive regulation	T, C	2847	618	987
Negative regulation	T, C	1062	196	379
TOTAL	-	8615	1795	3193

Table 1: BioNLP ST events, primary argument types and data statistics. Arguments abbreviate for (T)heme and (C)ause, with + marking arguments that can occur multiple times for an event. We refer to the task definition for details.

term "heterodimers". The SR data thus provides strong clues for the extraction of a Binding event involving both c-Rel and p50.

During the last few years, event extraction has gained much interest in the field of natural language processing (NLP) of biomedical text (Pyysalo et al., 2007; Kim et al., 2008; Kim et al., 2009). However, owing to the more complex nature of this task setting, performance rates are lower than for the extraction of simple binary relations. The currently best performing framework for event extraction obtains 53.29% F-score (Miwa et al., 2010), which is considerably lower than the performance reported for extraction of protein-protein interaction relations, ranging between 65% and 87% depending on the data set used for evaluation (Miwa et al., 2009).

In this paper, we will study how data on static relations can be applied to improve event extraction performance. First, we describe the various data sets (Section 2) and the text mining framework that was applied (Section 3). The main contributions of this paper are presented in Section 4, in which we study how static relation information can be integrated into an event extraction framework to enhance extraction performance. Finally, Section 5 presents the main conclusions of this work.

2 Data

In this section, we provide an overview of the two main data sets used in this work: event annotation (Section 2.1) and static relation annotation (Section 2.2).

2.1 Event Data

The BioNLP'09 Shared Task data, derived from the GENIA Event corpus (Kim et al., 2008), defines nine types of biomolecular events and is divided into three data sets: training data, development data and final test data, covering 800, 150 and 260 PubMed abstracts respectively. The event types and their statistics in the three data sets are shown in Table 1.

In the shared task setting, participants were provided with the gold annotations for Gene/Gene Product (GGP) named entities, and for all three data sets the texts of the abstracts and the gold GGP annotations are publicly available. However, while full gold event annotation is available for the training and development data sets, the shared task organizers have chosen not to release the gold annotation for the test data set. Instead, access to overall results for system predictions is provided through an online interface. This setup, adopted in part following a similar design by the organizers of the LLL challenge (Nédellec, 2005), is argued to reduce the possibility of overfitting to the test data and assure that evaluations are performed identically, thus maintaining comparability of results.

For the current study, involving detailed analysis of the interrelationships of two classes of annotations, the lack of access to the gold annotations of the test set rules this data set out as a potential target of study. Consequently, we exclude the blind test data set from consideration and use the development set as a test set.

To simplify the analysis, we further focus our efforts in this study on simple events involving only the given GGPs as participants. In the full shared task, events of the three Regulation types may take events as arguments, resulting in recursive event structures. These event types were found to be the most difficult to extract in the

SR type	Examples
term variant-of GGP	[RFX5 fusion protein], [Tax mutants], [I kappa B gamma isoforms]
term part-of GGP	[murine B29 promoter], [c-fos regulatory region], [transactivation domain] of Stat6, the nearby [J element] of the human DPA gene, the [consensus NF-kappa B binding site] of the E-selectin gene
GGP member-of term	The [Epstein-Barr virus oncoprotein] latent infection membrane protein 1, [Ikaros family members], PU.1 is a transcription factor belonging to the [Ets-family]
GGP subunit-of term	the [NF-kappa B complex] contains both RelA and p50, Human TAFII 105 is a cell type-specific [TFIID] subunit, [c-Rel/p65 heterodimers]

Table 2: Training examples of some of the SR types, including both noun phrase relations as well as relations between nominals. GGPs are underlined and terms are delimited by square brackets.

shared task evaluation (Kim et al., 2009). Furthermore, their inclusion introduces a number of complications for evaluation as well as analysis, as failure to extract a referenced event implies failure to extract events in which they appear as arguments. We note that even with the limitations of considering only the smallest of the three data sets and excluding Regulation events from consideration, the ST data still contains over 800 development test events for use in the analysis.

2.2 Static Relation Data

The data on relations is drawn from two recently introduced data sets. Both data sets cover specifically static relations where one of the participants is a GGP and the other a non-GGP term. The GGPs are drawn from the data introduced in (Ohta et al., 2009a) and the terms from the GENIA corpus term annotation (Kim et al., 2003), excluding GGPs. The first data set, introduced in (Pyysalo et al., 2009), covers static relations involving GENIA corpus terms that are annotated as participants in the events targeted in the BioNLP'09 shared task. The second data set, introduced in (Ohta et al., 2009b), contains annotation for relations holding between terms and GGPs embedded in those terms. In this study, we will use the non-embedded relations from the former data set, referring to this data as RBN for "Relations Between Nominals" in recognition of the similarity of the task setting represented by this data set and the task of learning semantic relations between nominals, as studied e.g. in SemEval (Girju et al., 2007; Hendrickx et al., 2009). We use all of the latter data set, below referred to as NPR for "Noun Phrase Relations". The NPR data set extends on the embedded part of the data introduced by (Pyysalo et al., 2009), increasing the coverage of terms included and the granularity of the annotated event types. While RBN only differentiates between a domain-specific *Variant* relation and four different part-whole relations, in NPR these are refined into more than 20 different types.

To apply these data sets together in a single framework, it was necessary to resolve the differences in the annotated relation types. First, as the finer-grained NPR types are organized in a hierarchy that includes the four part-whole relations of the RBN categorization as intermediate types (see Fig. 1 in Ohta et al. (2009b)), we collapsed the subtypes of each into these supertypes. While this removes some potentially useful distinctions, many of the finer-grained types are arguably unnecessarily detailed for the purposes of the event extraction task which, for example, makes no distinctions between events involving different gene components. Furthermore, the NPR annotations also define an Object-Variant class with multiple subtypes, but as these were judged too diverse to process uniformly, we did not collapse these subtypes as was done for part-whole relations. Rather, we divided them into "near" and "far" variants by a rough "functional distance" to the related GGP, as suggested by Ohta et al. (2009b). The relations *GGP-Modified Protein*, *GGP-Isoform* and *GGP-Mutant* were accepted into the "near" set, expected to provide positive features for inclusion in events, and the remaining subtypes into the "far" set, expected to provide negative indicators.

In addition to the primary annotation covering static relations, the RBN annotation only recognizes a mixed "other relation/out" category, used to annotate both GGP-term pairs for which the stated relation is not one of the targeted types (e.g. a causal relation) and pairs for which no relation is stated. Due to the heterogeneity of this category, it is difficult to make use of these annotations, and we have chosen not to consider them in this work.

By contrast, the NPR annotation also subdivides the "other relation" category into five specific types, providing an opportunity to also use the part of the data not strictly involving static relations. We judged the classes labeled *Functional*, *Experimental Method* and *Diagnosis and Therapeutics* to involve terms where contained GGP names are unlikely to be participants in stated events and thus provide features that could serve as potentially useful negative indicators for event extraction. As an example, the Functional category consists of GGP-term pairs such as *GGP inhibitor* and *GGP antibody*, where the term references an entity separate from the GGP, identified through a functional or causal relation to the GGP. As such terms occur in contexts similar to ones stating events involving the GGP, explicit marking of these cases could improve precision. Consider, for example, *GGP_1 binds GGP_2*, *GGP_1 binds GGP_2 promoter*, *GGP_1 binds GGP_2 inhibitor* and *GGP_1 binds GGP_2 antagonist*: a binding event involving *GGP_1* and *GGP_2* should be extracted for the first two statements but not the latter two.

Table 2 lists some interesting examples of static relation grouped by type, including both noun phrase relations as well as relations between nominals. The consolidated data combining the two static relations - related data sets are available at the GENIA project webpage.[1]

3 Methods

The text mining tool used for all analyses in this paper is based on the event extraction framework of Van Landeghem et al. (2009), which was designed specifically for participation in the BioNLP'09 Shared Task. In this framework, triggers are discovered in text by using automatically curated dictionaries. Subsequently, candidate events are formed by combining these triggers with an appropriate number of GGPs co-occurring in the same sentence. For each distinct event type, a classifier is then built using all training examples for that specific type. Final predictions are merged for all types, forming a complex interaction graph for each article in the test set.

To distinguish between positive instances and negatives, the framework extracts rich feature vectors by analyzing lexical and syntactic information from the training data. Subsequently, a support vector machine (SVM) is built with these training patterns. The patterns include trigrams, bag-of-word features, vertex walks and information about the event trigger. As part of the current study discusses the extension and generalization of these feature patterns (Section 4.4), we will briefly discuss the various types in this section.

To derive syntactic patterns, dependency parsing is applied using the Stanford parser (Klein and Manning, 2003; De Marneffe et al., 2006). Specifically, for each candidate event, the smallest subgraph is built including the relevant nodes for the trigger and the GGP names. Each edge in this subgraph then gives rise to a pattern including the information from the connecting nodes (or vertices) in combination with the syntactic relation specified by the edge. Trigger words and GGP names are blinded by replacing their text with the strings *protx* and *trigger* (respectively), resulting in highly general features.

Figure 3 depicts an exemplary dependency graph. For the Binding event between c-Rel and p50, the following vertex walks would be extracted: "trigger *nsubj* protx", "trigger *prep-as* heterodimer" and "heterodimer *prep-with* protx".

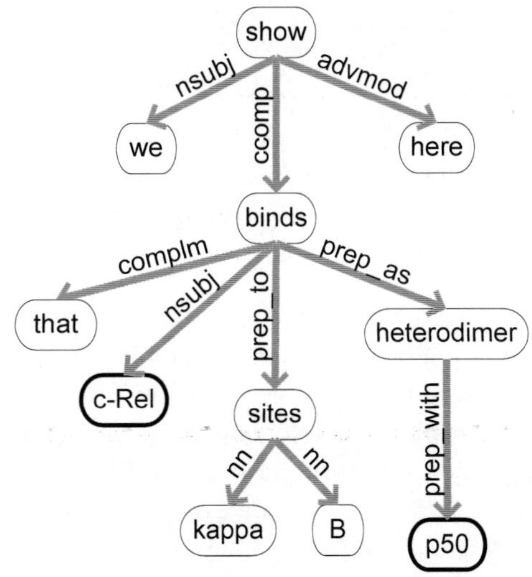

Figure 3: Dependency graph for the sentence "We show here that c-Rel binds to kappa B sites as heterodimers with p50". Words of the sentence form the nodes of the graph, while edges denote their syntactic dependencies.

[1]http://www-tsujii.is.s.u-tokyo.ac.jp/GENIA

Events	Training		Dev. test	
Pos. SR data	1190	32%	227	28%
Neg. SR data	841	22%	207	26%
All SR data	1635	44%	350	43%

Table 3: Number of events that can be linked to at least one static relation, including explicitly annotated "near miss" negative annotations, also showing percentage of all gold-standard events.

Furthermore, lexical information is provided by bag-of-word (BOW) features and trigrams. BOW features incorporate all words occurring as nodes in the dependency sub-graph. They include highly informative words such as "promoter". Trigrams are formed by combining three consecutive words in the sub-sentence delimited by the trigger and GGP offsets in text. They are capable of capturing common phrases such as "physical association with".

Finally, the lexical tokens of the event trigger are highly relevant to determine the plausibility of the event being a correct one. For example, "secretion" points to a Localization event, but more general words often lead to false candidate events, such as "present". The part of speech tags of the trigger words are also included as separate features.

During feature generation, all lexical patterns are stemmed using the Porter stemming algorithm (Porter, 1980), creating even more general features and reducing sparseness of the feature vectors.

4 Experiments

This section describes a thorough study on how data on static relations can be integrated into an event extraction framework. First, we will analyze the amount of useful complementary annotations across both data sets (Section 4.1). Next, we describe the generation and evaluation of new candidate events using terms involved in static relations, in an effort to boost recall of the event predictions (Section 4.2). To additionally improve on precision, we have implemented a false positive filter exploiting SR annotations of GGPs involved in relations judged to serve as negative indicators, such as "GGP inhibitor" (Section 4.3). Finally, Section 4.4 details experiments on the creation of more extensive features for event extraction by including static relation data.

	Predicted instances	Percentage of data set
Gene expression	63	17.70%
Transcription	34	41.46%
Protein catabolism	4	19.05%
Phosphorylation	20	42.55%
Localization	4	7.55%
Binding	73	29.44%
All events	198	24.54%

Table 4: Maximal recall performance of event instances involving at least one non-NE term as argument. These terms are functioning as aliases for the GGPs they are positively associated with.

4.1 Analysis of complementary data across the two data sets

To assess the usability of the SR data set for event extraction, we first analyze the amount of complementary annotations across the two data sets. On the document level, the static relations data contains some annotation for 87.6% of all training set articles and for 94.67% of the development test set, including both positive static relations as well as explicitly negated ones. Most articles from the event data set thus involve terms at least potentially involved in static relations.

Analyzing the overlap in more detail, we determined the number of events that could benefit from adding SR data by counting the number of events for which at least one GGP is also involved in a static relation (either a positive or a negative one). Table 3 shows the results of this evaluation. In the training data, 1635 events involve at least one GGP with SR annotation, which is 44% of all events in the gold-standard annotation. For the development test set, the number is 350 out of the 808 gold standard events, i.e. 43% of events. These development set events in particular will be the subject of this study.

4.2 Terms as aliases for related GGPs

Our first application of static relations in an event extraction framework involves the use of non-NE terms appearing in the SR data set as aliases for the GGPs they are positively associated with. In the event extraction framework, new candidate events can thus be formed by treating the terms as GGPs, and mapping them back to the real GGPs after classification. This procedure is motivated by the definition of the various SR types and the underlying biological processes. For example, if a complex is known to activate the expression of a cer-

	Recall	Precision	F-score
Gene expression	11.24%	81.63%	19.75%
Transcription	20.73%	89.47%	33.66%
Protein catabolism	19.05%	100.00%	32.00%
Phosphorylation	36.17%	100.00%	53.12%
Localization	3.77%	25.00%	6.56%
Binding	12.50%	45.59%	19.62%
All events	13.75%	67.27%	22.84%

Table 5: Performance of event instances involving at least one non-NE term as argument. These terms are functioning as aliases for the GGPs they are positively associated with.

	Baseline predictions	Merged predictions
Gene expression	77.01%	77.56%
Transcription	63.41%	64.24%
Protein catabolism	86.36%	86.36%
Phosphorylation	70.10%	76.47%
Localization	80.00%	76.77%
Binding	38.69%	40.52%
All events	64.71%	65.33%
All events (precision)	69.11%	67.19%
All events (recall)	60.84%	63.57%

Table 6: Performance of the event extraction framework. First column: using only normal events involving GGPs ("baseline"). Second column: merging the new predictions (Table 5) with the first ones. All performance rates indicate F-score, except for the last two rows.

tain target GGP, then the various subunits of this complex can be annotated as participants in that event.

Obviously, this approach has some intrinsic limitations as not all GGPs occurring as arguments in events have a corresponding term that could be used as alias. However, from Table 3 it is clear that it should still be possible to extract 227 gold standard cases. To test the limitation, we have used the event extraction framework detailed in Section 3, removing the SVM classifier from the pipeline and simply labeling all candidate events as positive predictions. The result indicates that the framework is capable of retrieving 198 of the 227 gold standard cases (Table 4). The 29 missing events are due to trigger words not appearing (frequently) in the training set and thus missing from the dictionary, preventing the event to be formed as a candidate in the framework.

Our results thus show that nearly 25% of all events are potentially retrievable by using non-NE terms as aliases for GGPs. However, the analysis also indicates that in this approach, some event types are much easier to extract than others. For example, less than 8% of Localization events can be found with this setup, while maximal recall for Phosphorylation events is over 40%. These results reflect the intrinsic differences between event types and the ways in which they are typically expressed, and suggest that it should be beneficial for event extraction to take these differences into account when incorporating static relations.

Having established an upper bound for recall, a subsequent experiment involves treating the newly created instances as normal candidate events. For classification, we use an SVM trained on regular candidate events involving GGPs, as this ensures sufficient training material.

Both lexical and syntactic patterns are expected to be similar for events involving either non-NE terms or GGPs. To test this hypothesis, we have run the event-extraction pipeline for these new instances. Evaluation is performed with the standard evaluation script provided by the BioNLP'09 Shared Task organizers, which measures the percentage of true events amongst all predictions (precision), the percentage of gold-standard events recovered (recall) and the harmonic mean of these two metrics (F-score). The results are detailed in Table 5. While we have already established that recall is subject to severe limitations (Table 4), we note in particular the high precision rates of the predictions. In particular, four out of six event types achieve a precision rate higher than 80%.

To allow for a meaningful comparison, these results should be put into perspective by merging the new predictions with the predictions of a baseline extractor and comparing against this baseline (Table 6). This analysis reveals interesting results: while overall performance increases slightly from 64.71% to 65.33% F-score, this trend is not common to all event types. For instance, prediction of Localization drops 3.23% points F-score. Considering the maximum recall results, this is not entirely surprising and confirms the hypothesis that the prediction of Localization events will not benefit from static relation data in this approach.

However, we do observe a considerable increase in performance for Phosphorylation (6.37% points F-score) events and some increase for Binding events (1.83% points F-score). This performance boost is mainly caused by an increase in recall (10.64% and 4.43% points, respectively). When considering all protein events, recall is increased from 60.84% to 63.57% (Table 6, last row). These results clearly indicate that the inclusion of static relations can improve recall while retaining and even slightly improving general performance.

4.3 Using static relations to filter false positive events

To further improve event extraction performance, we have designed a false-positive (FP) filter using specific categories of relations serving as negative indicators for event extraction. In particular, we have used the "far variants" and *Functional* relation annotations, as described in Section 2.2. For each such relation, we add the GGP involved to the FP filter, as the GGP should not participate in any event. Thus, for example, the GGP in "GGP antibodies" would be filtered as the GGP is considered too far removed from the containing term to be a participant in any event in the context.

In the development test set, this strategy has automatically identified 24 relevant GGP mentions that should not be annotated as being involved in any event. Even though this number is relatively small, we aim at designing a high specificity FP filter while relying on the SVM classifier to solve more ambiguous cases.

Applying the FP filter to the baseline result detailed in Table 6, we find that 3 events are discarded from the set of predictions. All three instances represented false positives; two of them were Binding events and one a Gene expression event. Overall precision and F-score increased by 0.30% points and 0.13% points, respectively.

4.4 Extended feature representation incorporating information on static relations

The last type of experiment aims to boost both precision and recall by substantially extending the feature generation module for event extraction using the newly introduced SR data. Table 3 shows that such an enhanced feature representation could influence 1190 events in the training data (1635 events including negative annotations) and 227 events in the development test data (350 including negative), covering a significant part of the data set.

Building further on the feature generation module described in Section 3, we have added a range of new features to the feature vectors while also providing enhanced generalization of existing features. Generalization is crucial for the text mining framework as it enables the extraction of relations from new contexts and forms of statements.

First, for each term involved in a static relation with a GGP, the string of the term is included as a separate feature. This generates relation-associated features such as "tyrosine", which is strongly correlated with Phosphorylation events. For terms spanning multiple tokens, we additionally include each token as a separate feature, capturing commonly used words such as "promoter" or "receptor". Each distinct feature is linked to its specific relation type, such as Part-of or Member-collection (Section 2.2). To make use of annotation for "near-miss" negative cases, we generate features also for these relations, marking each feature to identify whether it was derived from a positive or negative annotation.

Additionally, we introduced a new feature type expressing whether or not the trigger of the event is equal to a term related to one or more GGPs involved in the event. As an example, suppose the candidate event is triggered by the word "homodimer". If the GGP involved is annotated as being a subunit of this homodimer, this provides a strong clue for a positive event. Similarly, the explicit negation of the existence of any static relation indicates a negative event.

Apart from these new features, we have also investigated the use of static relations to create more general lexical patterns. In particular, we have adjusted the lexical information in the feature vector by blinding terms involved in relevant relations, depending on the specific type of relation. For each such term, the whole term string is replaced by one word, expressing the type of the static relation and whether the relation is positive or negative. This results in more general patterns such as "inhibit prep-to partx" (vertex walk) or "activ in nonpartx" (trigram). In Figure 3, "heterodimer" would be blinded as "complexx" as both c-Rel and p50 are members of this complex.

Initial experiments with the extended feature representation showed that an increase in performance could be obtained on the development test set, achieving 61.34% recall, 69.58% precision and 65.20% F-score. However, it also became clear that not all event types benefit from the new features. Surprisingly, Binding is one such example. We hypothesize that this is mainly due to the intrinsic complexity of Binding events, requiring an even more advanced feature representation.

	Baseline predictions	New predictions
Gene expression	77.01%	78.06%
Transcription	63.41%	63.80%
Protein catabolism	86.36%	86.36%
Phosphorylation	70.10%	76.29%
Localization	80.00%	84.21%
Binding	38.69%	38.34%
All events	64.71%	65.73%
All events (precision)	69.11%	69.99%
All events (recall)	60.84%	61.96%

Table 7: Performance of the event extraction framework. First column: using the baseline feature representation. Second column: using the extended feature representation. All performance rates indicate F-score, except for the last two rows.

To take the inherent differences between various event types into account, we selected the optimal set of features for each type. In a new experiment, the feature generation step thus depends on the event type under consideration. Table 7 details the results of this optimization: an overall F-score of 65.73% is achieved. Similar to the experiments in Section 4.2, the F-score for the prediction of Phosphorylation events increases by 6.19% points. Additionally, in this experiment we obtain an increase of 4.21% points in F-score for Localization events, even though we were unable to improve on them when using terms as aliases for additional candidate events (Section 4.2). Additional experiments suggested the reason to be that while the Localization event type in general does not benefit from positive static relations, negative static relations seem to provide strong clues to the SVM classifier.

5 Conclusion

We have presented the first study on the applicability of static relations for event prediction in biomedical texts. While data on static relations can offer a more detailed representation of biomolecular events, it can also help to boost the performance of event prediction. We have performed three sets of experiments to investigate these opportunities. First, we have designed new candidate events by treating non-NE terms as aliases for the GGPs they are associated with. By augmenting the normal event predictions with predictions for these new candidates, we have established a considerable increase in recall. Next, we have implemented a false positive filter to improve precision, by exploiting annotation for relations judged to imply only distant associations of the GGP and the enclosing term. Finally, the last type of experiment involves integrating complementary data on static relations to obtain more informative feature vectors for candidate events. Results show that both recall and precision can be increased slightly by this last, more complex configuration.

During the experiments, it has become clear that there are important differences between the data sets of distinct event types. For example, we have found that Phosphorylation events benefit the most from added static relations data, while Localization events can be enhanced using only features of negative static relation annotations. For some event types, such as Protein catabolism, the current techniques for integration of static relations do not generate a performance boost. However, our findings pave the way for experiments involving more detailed representations, taking the intrinsic properties of the various event types into account and combining the various ways of integrating the new information. We regard these opportunities as promising future work.

Finally, having established the potential added value offered by data on static relations in an event extraction framework, additional future work will focus on the automatic extraction of the static relations. Similar relations have been considered in numerous recent studies, and while challenges to reliable prediction remain, several methods with promising performance have been proposed (Girju et al., 2007; Hendrickx et al., 2009). By integrating predictions from both static relations and events instead of using gold standard relation annotations, we will be able to study the effect of the relation information on new data, including the shared task test set. Such experiments are key to establishing the practical value of static relations for biomolecular event extraction.

Acknowledgments

SVL would like to thank the Research Foundation Flanders (FWO) for funding her research. The work of SP and TO was partially supported by Grant-in-Aid for Specially Promoted Research (MEXT, Japan).

References

M. De Marneffe, B. Maccartney, and C. Manning. 2006. Generating typed dependency parses from phrase structure parses. In *Proceedings of LREC-06*, pages 449–454.

Roxana Girju, Preslav Nakov, Vivi Nastase, Stan Szpakowicz, Peter Turney, and Deniz Yuret. 2007. Semeval-2007 task 04: Classification of semantic relations between nominals. In *Proceedings of the Fourth International Workshop on Semantic Evaluations (SemEval-2007)*, pages 13–18, Prague, Czech Republic, June. Association for Computational Linguistics.

Iris Hendrickx, Su Nam Kim, Zornitsa Kozareva, Preslav Nakov, Diarmuid Ó Séaghdha, Sebastian Padó, Marco Pennacchiotti, Lorenza Romano, and Stan Szpakowicz. 2009. Semeval-2010 task 8: Multi-way classification of semantic relations between pairs of nominals. In *Proceedings of the Workshop on Semantic Evaluations: Recent Achievements and Future Directions (SEW-2009)*, pages 94–99, Boulder, Colorado, June. Association for Computational Linguistics.

Jin-Dong Kim, Tomoko Ohta, Yuka Tateisi, and Jun'ichi Tsujii. 2003. GENIA corpus - a semantically annotated corpus for bio-textmining. *Bioinformatics*, 19(suppl. 1):i180–i182.

Jin-Dong Kim, Tomoko Ohta, and Jun'ichi Tsujii. 2008. Corpus annotation for mining biomedical events from literature. *BMC Bioinformatics*, 9(1):10.

Jin-Dong Kim, Tomoko Ohta, Sampo Pyysalo, Yoshinobu Kano, and Jun'ichi Tsujii. 2009. Overview of bionlp'09 shared task on event extraction. In *Proceedings of the BioNLP 2009 Workshop Companion Volume for Shared Task*, pages 1–9, Boulder, Colorado, June. Association for Computational Linguistics.

Dan Klein and Christopher D. Manning. 2003. Accurate unlexicalized parsing. In *Proceedings of the 41st Annual Meeting of the Association for Computational Linguistics*, pages 423–430, Sapporo, Japan, July. Association for Computational Linguistics.

Makoto Miwa, Rune Saetre, Yusuke Miyao, and Jun'ichi Tsujii. 2009. A rich feature vector for protein-protein interaction extraction from multiple corpora. In *EMNLP '09: Proceedings of the 2009 Conference on Empirical Methods in Natural Language Processing*, pages 121–130, Morristown, NJ, USA. Association for Computational Linguistics.

Makoto Miwa, Rune Saetre, Jin-Dong D. Kim, and Jun'ichi Tsujii. 2010. Event extraction with complex event classification using rich features. *Journal of bioinformatics and computational biology*, 8(1):131–146, February.

Claire Nédellec. 2005. Learning language in logic - genic interaction extraction challenge. In *Proceedings of the Learning Language in Logic Workshop (LLL'05)*.

Tomoko Ohta, Jin-Dong Kim, Sampo Pyysalo, Yue Wang, and Jun'ichi Tsujii. 2009a. Incorporating genetag-style annotation to genia corpus. In *Proceedings of the BioNLP 2009 Workshop*, pages 106–107, Boulder, Colorado, June. Association for Computational Linguistics.

Tomoko Ohta, Sampo Pyysalo, Kim Jin-Dong, and Jun'ichi Tsujii. 2009b. A re-evaluation of biomedical named entity - term relations. In *Proceedings of LBM'09*.

M. F. Porter. 1980. An algorithm for suffix stripping. *Program*, 14(3):130–137.

Sampo Pyysalo, Filip Ginter, Juho Heimonen, Jari Björne, Jorma Boberg, Jouni Järvinen, and Tapio Salakoski. 2007. Bioinfer: a corpus for information extraction in the biomedical domain. *BMC bioinformatics*, 8(1):50+.

Sampo Pyysalo, Tomoko Ohta, Jin-Dong Kim, and Jun'ichi Tsujii. 2009. Static relations: a piece in the biomedical information extraction puzzle. In *Proceedings of the BioNLP 2009 Workshop*, pages 1–9, Boulder, Colorado, June. Association for Computational Linguistics.

Sofie Van Landeghem, Yvan Saeys, Bernard De Baets, and Yves Van de Peer. 2009. Analyzing text in search of bio-molecular events: a high-precision machine learning framework. In *BioNLP '09: Proceedings of the Workshop on BioNLP*, pages 128–136, Morristown, NJ, USA. Association for Computational Linguistics.

Author Index

Ananiadou, Sophia, 132
Apostolova, Emilia, 81

Bergman, Casey M., 72
Björne, Jari, 28, 108

Cho, Han-Cheol, 132
Choi, Jinwook, 130
Chowdhury, Md. Faisal Mahbub, 83

Díaz, Alberto, 55

Elhadad, Noemie, 64

Fiszman, Marcelo, 46
Frunza, Oana, 91

Gerner, Martin, 72
Ginter, Filip, 28
Gonzalez, Graciela, 117
Guo, Yufan, 99

Hara, Tadayoshi, 37
Harshfield, Benjamin, 141
Heimonen, Juho, 108
Huang, Minlie, 10

Inkpen, Diana, 91

Jha, Mukund, 64

Kilicoglu, Halil, 46
Kim, Jin-Dong, 19
Korhonen, Anna, 99

Lavelli, Alberto, 83
Leaman, Robert, 117
Liakata, Maria, 99
Liu, Jingchen, 10

Mao, Chunhong, 132
Marimpietri, Sean, 46
Miwa, Makoto, 19, 37
Morante, Roser, 126

Naderi, Nona, 128
Nenadic, Goran, 72

Ohta, Tomoko, 19, 132, 144

Park, Heekyong, 130
Plaza, Laura, 55
Pyysalo, Sampo, 19, 28, 37, 132, 144

Rindflesch, Thomas, 46
Rosemblat, Graciela, 46

Salakoski, Tapio, 28, 108
Silins, Ilona, 99
Skariah, Annie, 117
Sobral, Bruno, 132
Stenius, Ulla, 99
Stevenson, Mark, 55
Stubbs, Amber, 141
Sullivan, Dan, 132
Sullivan, Ryan, 117
Sun, Lin, 99

Tomuro, Noriko, 81
Tsujii, Jun'ichi, 19, 28, 37, 132

Van de Peer, Yves, 144
Van Landeghem, Sofie, 144
Vlachos, Andreas, 1

Witte, René, 128
Wojtulewicz, Laura, 117

Yang, Jian, 117

Zhu, Xiaoyan, 10

Association for Computational Linguistics
209 N. Eighth Street
Stroudsburg, Pennsylvania 18360